AF581801

Ozone Evolution in the Past and Future

Ozone Evolution in the Past and Future

Editor

Eugene Rozanov

MDPI • Basel • Beijing • Wuhan • Barcelona • Belgrade • Manchester • Tokyo • Cluj • Tianjin

Editor
Eugene Rozanov
Physikalisch-Meteorologisches Observatorium
Davos World Radiation Centre
Switzerland

Editorial Office
MDPI
St. Alban-Anlage 66
4052 Basel, Switzerland

This is a reprint of articles from the Special Issue published online in the open access journal *Atmosphere* (ISSN 2073-4433) (available at: https://www.mdpi.com/journal/atmosphere/special_issues/Ozone).

For citation purposes, cite each article independently as indicated on the article page online and as indicated below:

LastName, A.A.; LastName, B.B.; LastName, C.C. Article Title. *Journal Name* **Year**, *Article Number*, Page Range.

ISBN 978-3-03936-828-0 (Hbk)
ISBN 978-3-03936-829-7 (PDF)

Contents

About the Editor

Eugene Rozanov (Dr.) Eugene Rozanov was born in 1955 in Saint Petersburg, Russia. He studied physics at the State University of Saint Petersburg. In 1979, he started working in the Main Geophysical Observatory and, in 1986, received his Ph.D. for his work in atmospheric physics. In 1997, he moved to the University of Illinois in Urbana-Champaign to work on chemistry-climate model development. In 2001, he started his work on the solar variability influence on climate and the ozone layer in IAC ETH and PMOD/WRC, Switzerland. He participated in the preparation of the IPCC-2007 report and WMO and SPARC international scientific assessments on the development of the stratospheric ozone layer. In 2013, he created and led a climate research group in PMOD/WRC and, since that time, has been involved in several projects aimed at the study of the ozone layer and climate evolution in the past and future.

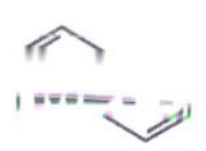

Editorial

Preface: Ozone Evolution in the Past and Future

Eugene Rozanov [1,2]

1 Physikalisch-Meteorologisches Observatorium Davos World Radiation Centre, 7260 Davos, Switzerland; eugene.rozanov@pmodwrc.ch

2 Institute for Atmospheric and Climate Science, Swiss Federal Institute of Technology Zurich, 8092 Zurich, Switzerland

Received: 24 June 2020; Accepted: 2 July 2020; Published: 3 July 2020

Keywords: ozone layer evolution; modeling; ultraviolet radiation; forcing

---

1. Introduction

The stratospheric ozone plays an important role in the protection of the biosphere from the dangerous ultraviolet radiation of the sun. It also forms the temperature structure of the stratosphere and therefore, has a direct influence on the general circulation and the surface climate [1,2]. The tropospheric ozone can damage the biosphere, impact human health, and plays a role as a powerful greenhouse gas [3,4]. That is why the understanding of the past and future evolution of the ozone in different atmospheric layers, its influence on surface UV radiation doses, and human health is important. The discovery of the ozone hole in 1984/1985 led to limitations in the production of halogen-containing, ozone-depleting substances by the Montreal Protocol and its amendments (MPA). This measure aimed to prevent ozone layer depletion and to guarantee future recovery of the ozone shield in the future. However, the expected ozone recovery and the effectiveness of MPA are now being questioned due to the continuous negative ozone trend in the lower stratosphere [5] and the appearance of a large ozone hole over the Arctic in 2020 [6]. The abovementioned facts inspired the appearance of this Special Issue. Initially, I aimed to publish studies of the stratospheric and total column ozone trends, but during work on this issue, I decided to also accept papers related to tropospheric ozone effects and trends. Thus, this Special Issue includes recent experimental, statistical, and modeling works describing several aspects of the ozone layer's state and evolution. In the next section, the individual contributions are summarized and divided into two broad topics covering the issues related to the stratospheric and tropospheric ozone layer.

2. Summary of This Special Issue

2.1. Stratospheric Ozone Evolution

Smyshlyaev et al. [7] attempted to elucidate the contribution of different natural and anthropogenic factors to the observed stratospheric ozone variability for the 1979–2020 period, exploiting their state-of-the-art chemistry–climate model. They also compared results from the late twentieth and early twenty-first century against ground-based and satellite observations. In particular, their model showed that WMO (World Meteorological Organization) 2011 scenarios probably overestimated the reduction in hODS (halogen containing Ozone Depleting Substances) emissions into the atmosphere in 2010–2020. Egorova et al. [8] addressed ozone layer evolution during the early 20th century using their chemistry–climate model driven by all known external drivers. They concluded that enhanced solar UV radiation is the main driver of the stratospheric and total column ozone changes. This finding would be useful to constrain solar forcing magnitude if some reconstructions of the ozone are available. Chubarova et al. [9] analyzed the evolution of erythemal radiation over Northern Eurasia caused by total column ozone and cloud amount variabilities, acquired from the observations and model simulations.

They found generally positive trends for all months and locations except in the central Arctic in July, which is driven by the increase in cloud amount. These conclusions are in agreement with a recently published modeling study [10] on future ozone layer and UV evolutions. The accuracy of the trend analysis depends on the quality of the observational time series. Krizan et al. [11] carefully investigated the ozone time series from MERRA-2 reanalysis and found several discontinuities, which can affect the ozone trend analysis mostly above the 4 hectopascal pressure layer. A new and promising statistical method for the trend analysis based on Empirical Wavelet Transform was proposed and exploited by Mbatha and Bencherif [12]. Using this approach, they performed a detailed analysis of the total column ozone data measured in Buenos Aires, Argentina, and obtained a robust ozone recovery during the 2010–2017 period, excluding ozone decline in 2015 caused by the Calbuco volcanic event. Mkololo et al. [13] studied stratospheric ozone changes and stratosphere–troposphere exchange events over the Irene station during the 2000–2007 and 2012–2015 periods. They obtained a negative ozone trend by up to 9% in the 12–24 km layer in the 2012–2015 period relative to the 2000–2003 average. This result is in agreement with the analysis of the satellite data by Ball et al. [5].

2.2. *Tropospheric Ozone Evolution*

Tropospheric ozone layer evolution during the early 20th century was addressed by Egorova et al. [8], using their chemistry–climate model driven by anthropogenic and natural forcing. They obtained an up to 20% enhancement of the ozone in the northern boundary layer during the 1910–1940 period, driven by anthropogenic ozone precursors. Diaz et al. [14] evaluated the 30-year surface ozone trend in the United Kingdom using measurements from 13 rural and 6 urban sites. They showed that introduced emission control facilitated cleaner air with a much smaller number of days with high surface ozone abundance. They also stated that the highest number of days with high surface ozone concentration is reached in May for all stations. The contribution of the stratospheric ozone intrusion to the events with high surface ozone concentration in China was analyzed by Wang et al. [15] using a Lagrangian Particle Dispersion Model driven by meteorological reanalysis data. They concluded that stratospheric intrusions can be responsible for about 15% of severe ozone pollution events and should be considered in surface ozone forecast and control. The potential impact of such strong pollution events on mortality rate was evaluated by Park [16]. He showed that in Seoul (South Korea), elevated surface ozone substantially contributes to mortality rates during spring, while the high air temperature is more dangerous during the summer. These conclusions are useful for local authorities and air quality control system developers.

3. Conclusions

This Special Issue consists of nine contributions discussing different aspects of ozone layer evolution. I hope the obtained results will be of interest to the ozone community and inspire new studies in this direction using models and observations.

Funding: This work was supported by the Swiss National Science Foundation under 200020_182239 (POLE).

Acknowledgments: I would like to thank all the contributors to this Special Issue, as well as the Editorial team of *Atmosphere.*

Conflicts of Interest: The author declares no conflict of interest.

References

1. Bais, A.F.; Lucas, R.M.; Bornman, J.F.; Williamson, C.E.; Sulzberger, B.; Austin, A.T.; Wilson, S.R.; Andrady, A.L.; Bernhard, G.; Aucamp, P.J.; et al. Environmental effects of ozone depletion, UV radiation, and interactions with climate change: UNEP Environmental Effects Assessment Panel, update 2019. *Photochem. Photobiol. Sci.* **2020**, *19*, 542–584. [CrossRef]
2. World Meteorological Organization (WMO). *Scientific Assessment of Ozone Depletion: 2018, Global Ozone Research and Monitoring Project—Report No. 58*; WMO: Geneva, Switzerland, 2018; 588p.

3. Fowler, D.; Amann, M.; Anderson, R.; Ashmore, M.; Cox, P.; Depledge, M.; Derwent, D.; Grennfelt, P.; Hewitt, N.; Jenkin, M.; et al. *Ground-Level Ozone in the 21st Century: Future Trends, Impacts, and Policy Implications*; Royal Society Policy Document 15/08; The Royal Society: London, UK, 2008.
4. IPCC. *Climate Change 2013: The Physical Science Basis. Contribution of Working Group I to the Fifth Assessment Report of the Intergovernmental Panel on Climate Change*; Stocker, T.F., Qin, D., Plattner, G.-K., Tignor, M., Allen, S.K., Boschung, J., Nauels, A., Xia, Y., Bex, V., Midgley, P.M., Eds.; Cambridge University Press: Cambridge, UK; New York, NY, USA, 2013; 1535p.
5. Ball, W.T.; Alsing, J.; Mortlock, D.J.; Staehelin, J.; Haigh, J.D.; Thomas, P.; Tummon, F.; Stübi, R.; Stenke, A.; Bourassa, A.; et al. Continuous decline in lower stratospheric ozone offsets ozone layer recovery. *Atmos. Chem. Phys.* **2018**, *18*, 1379–1394. [CrossRef]
6. Witze, A. Rare ozone hole opens over Arctic—And it's big. *Nature* **2020**, *580*, 18–19. [CrossRef] [PubMed]
7. Smyshlyaev, S.; Galin, V.; Blakitnaya, P.; Jakovlev, A. Numerical modeling of the natural and manmade factors influencing past and current changes in polar, mid-latitude and tropical ozone. *Atmosphere* **2020**, *11*, 76. [CrossRef]
8. Egorova, T.; Rozanov, E.; Arsenovic, P.; Sukhodolov, T. Ozone layer evolution in the early 20th century. *Atmosphere* **2020**, *11*, 169. [CrossRef]
9. Chubarova, N.; Pastukhova, A.; Zhdanova, E.; Volpert, E.; Smyshlyaev, S.; Galin, V. Effects of ozone and clouds on temporal variability of surface UV radiation and UV resources over northern eurasia derived from measurements and modeling. *Atmosphere* **2020**, *11*, 59. [CrossRef]
10. Eleftheratos, K.; Kapsomenakis, J.; Zerefos, C.S.; Bais, A.F.; Fountoulakis, I.; Dameris, M.; Jöckel, P.; Haslerud, A.S.; Godin-Beekmann, S.; Petropavlovskikh, I.; et al. Possible effects of greenhouse gases to ozone profiles and DNA active UV-B irradiance at ground level. *Atmosphere* **2020**, *11*, 228. [CrossRef]
11. Krizan, P.; Kozubek, M.; Lastovicka, J. Discontinuities in the ozone concentration time series from MERRA 2 reanalysis. *Atmosphere* **2019**, *10*, 812. [CrossRef]
12. Mbatha, N.; Bencherif, H. Time series analysis and forecasting using a novel hybrid LSTM data-driven model based on empirical wavelet transform applied to total column of ozone at buenos aires, argentina (1966–2017). *Atmosphere* **2020**, *11*, 457. [CrossRef]
13. Mkololo, T.; Mbatha, N.; Sivakumar, V.; Bègue, N.; Coetzee, G.; Labuschagne, C. Stratosphere–Troposphere exchange and O_3 variability in the lower stratosphere and upper troposphere over the irene SHADOZ site, South Africa. *Atmosphere* **2020**, *11*, 586. [CrossRef]
14. Diaz, F.; Khan, M.; Shallcross, B.; Shallcross, E.; Vogt, U.; Shallcross, D. Ozone trends in the United Kingdom over the last 30 years. *Atmosphere* **2020**, *11*, 534. [CrossRef]
15. Wang, Y.; Wang, H.; Wang, W. A stratospheric intrusion-influenced ozone pollution episode associated with an intense horizontal-trough event. *Atmosphere* **2020**, *11*, 164. [CrossRef]
16. Park, S. Assessing the impact of ozone and particulate matter on mortality rate from respiratory disease in Seoul, Korea. *Atmosphere* **2019**, *10*, 685. [CrossRef]

Article

Numerical Modeling of the Natural and Manmade Factors Influencing Past and Current Changes in Polar, Mid-Latitude and Tropical Ozone

Sergei P. Smyshlyaev [1,*], Vener Y. Galin [2], Polina A. Blakitnaya [1] and Andrei R. Jakovlev [1]

1 Russian State Hydrometeorological University, Voronezhskaya Str., 79, 192007 Saint-Petersburg, Russia; pzim@rshu.ru (P.A.B.); endrusj@rambler.ru (A.R.J.)
2 Institute of Numerical Mathematics RAS, Gubkina str., 8, 119991 Moscow, Russia; venergalin@yandex.ru
* Correspondence: smyshl@rshu.ru

Received: 31 October 2019; Accepted: 28 December 2019; Published: 8 January 2020

Abstract: A chemistry–climate model of the lower and middle atmosphere is used to compare the role of natural and anthropogenic factors in the observed variability of stratospheric ozone. Numerical experiments have been carried out on several scenarios of separate and combined effects of solar activity, stratospheric aerosol, sea surface temperature, greenhouse gases, and ozone-depleting substances emissions on ozone for the period from 1979 to 2020. Simulations for the past and present periods are compared to the results of ground-based and satellite observations. Estimates of observed trends in column total ozone for the entire period 1980–2018 and separately for the late twentieth and early twenty-first century are presented.

Keywords: stratospheric ozone; natural and anthropogenic factors; numerical modeling; satellite observations; trend estimations

---

1. Introduction

The change in the content of atmospheric ozone, which protects life on Earth from the harmful effects of the hard part of the ultraviolet radiation of the Sun, has attracted the attention of scientists and policymakers for many years in connection with the observed negative trends in ozone global content since the early 1980s [1–5]. In this regard, attention has concentrated on the impact of anthropogenic factors associated with industrial and household emissions of chlorine and bromine gases (ozone-depleting species—ODS) into the atmosphere, contributing to the destruction of the ozone layer. ODS are inert in the lower atmosphere, but, while rising into the stratosphere, they fall under the influence of intense fluxes of ultraviolet (UV) solar radiation. Thus, they are destroyed and release a large amount of chlorine and bromine radicals that destroy ozone molecules in catalytic cycles [6,7]. The rapid and timely adoption of measures to limit ODS emissions under the Montreal Protocol and its subsequent amendments has halted the growth of ODS in the stratosphere, and the stratospheric ozone is now expected to begin to recover to early 1980s levels [8,9].

Studies of trends in total ozone in the late twentieth and early twenty-first centuries depicted that they were negative until the mid-90s and in subsequent years showed signs of a weak, mostly insignificant positive trend [1–5,8,9]. At the same time, significant trends were observed in the upper and lower stratosphere, whereas no changes were observed in the middle stratosphere, where the largest amount of ozone is located [10]. At the same time, trends in different latitudinal regions differ quantitatively [11,12]. The magnitude, significance, and even sign of the trend over the last period can change significantly with each additional year of observation, so despite previously published results of ozone trend estimates, new results that take into account recent measurements may be useful to the scientific community.

Meanwhile, the results of measurements show that, despite the measures taken to limit the impact of anthropogenic factors on ozone, its recovery has not been as fast as expected [11–14], and in some regions in the lower stratosphere there still is a slight decrease [15]. These facts make timely and relevant the detailed study of the influence of natural factors such as solar activity, stratospheric aerosol content, temperature of the atmosphere and ocean, the area covered by ocean ice, and the general circulation of the atmosphere on the content of stratospheric ozone. The combined effects of natural and anthropogenic factors can lead to non-linear effects in ozone variability, weighed down, in addition, by a large number of positive and negative feedbacks superimposed on the previously expected faster recovery of the ozone layer as a result of a reduction in ODS [16,17].

To study all the features of nonlinear interactions between physical and chemical processes in different altitude layers of the atmosphere that affect the change of stratospheric ozone content under the influence of natural and anthropogenic factors, the most rational approach utilizes numerical modeling of chemical and climatic processes in the atmosphere, taking into account their interaction and feedbacks, and comparing the results of numerical modeling to the results of observations [18,19]. This approach is proposed in this paper, combining the analysis of satellite observations of total ozone from the early 80s to the present time with numerical modeling of atmospheric ozone variability from 1980 to 2020, taking into account the influence of chemical and dynamical factors of ozone variability as well as their interaction.

The purpose of this work is to study the trends of total ozone in different latitudinal regions during different periods of time from the early 80s of the twentieth century to the end of the 10s of the twenty-first century to study the influence of natural, anthropogenic, physical, and chemical factors on the observed variability of total ozone in different latitudinal regions and to highlight the predominant role of individual factors in individual time periods.

2. Methodology

To study the interannual variability of the total column ozone in the tropical, middle, and polar latitudes during 1980–2020, the results of satellite observations and numerical modeling using a chemistry–climate model (CCM) were analyzed. Satellite observations of solar backscatter ultraviolet (SBUV), total ozone mapping spectrometer (TOMS), and ozone monitoring instrument (OMI) as well as the results of calculations using a CCM of the Institute of Numerical Mathematics and the Russian State Hydrometeorological University (INM RAS–RSHU CCM), were averaged for each year and the latitude bands 90S–60S (Antarctica), 60S–30S (middle latitude southern hemisphere), 30S–30N (tropics), 30N–60N (middle latitude northern hemisphere), and 60N–90N (the Arctic).

The solar backscattering ultraviolet (SBUV) radiometer [20] is a series of operational remote sensors on Nimbus and NOAA weather satellites in solar synchronous orbits that provide measurements of total ozone with global coverage in the stratosphere, as well as ozone profiles, from 1970 to 2017 [21]. Data from different satellites are calibrated among themselves, resulting in a merged set of combined data [9], continuously covering the period from 1979 to 2017.

The Total Ozone Mapping Spectrometer (TOMS) is a NASA satellite instrument for measuring ozone concentration [22–25]. Of the five instruments built, four entered successful orbit on the Nimbus-7, Meteor-3, Advanced Earth Observing Satellite (ADEOS), and Earth Probe (EP) satellites. Instruments on the Nimbus-7 and Meteor-3 satellites produced a continuous set of data on the total ozone content from November 1978 to December 1994 [23,24]. The instrument on the ADEOS satellite gave data from 1996 to mid-1997, but did not cover a full year. Since mid-1996, when the Earth Probe satellite was launched [25], the TOMS instrument on that satellite transmitted data on the total ozone content until 2006. As a continuation and replacement of the TOMS series of instruments, an ozone monitoring instrument (OMI) was launched on the Aura satellite in 2004, which provides data on the total ozone content to date [26].

Due to the fact that only SBUV satellite measurements provide consistent data on the total ozone content for the entire study period from 1980 to 2018, they were used in this work to study trends in ozone content, both for the entire period and for individual time intervals. TOMS and OMI data that do not allow for a consistent series for the entire study period were considered complementary to the SBUV data to clarify trends in individual time periods. For the average annual SBUV data values in each latitudinal region, linear trends of total ozone were calculated using linear regression formulas [27] for the entire analyzed period from 1980 to 2020 and separately for the end of the twentieth (1980–2000) and the beginning of the twenty-first (2000–2017) centuries. As a quantitative characteristic of the trend, the linear regression coefficient characterizing the rate of change in the total ozone content (Dobson units per year) was calculated. For the end of the twentieth century, when there was a sharp transition on a global scale from a stable decrease in ozone from 1980 to the mid-90s to a sharp increase from the mid-90s to the early 2000s, separate trends were estimated for these two multidirectional periods of variability in total ozone for different latitudinal regions. For all calculated trends, their statistical significance was estimated by Fisher's criterion [27] by comparing the coefficient of the linear trend with the spread of data relative to the regression line, and then the probability density function (PDF) and root mean square (RMS) error were evaluated.

To assess the role of various factors affecting the interannual variability of ozone content, numerical experiments were performed with the INM RAS–RSHU CCM (version 1) [28]. The model has a spatial resolution of 4 × 5 degrees latitude/longitude and covers the altitude range from the Earth's surface up to the mesopause (around 90 km) with 39 sigma-pressure levels with variable pressure spacing [28]. In order to take into account an interaction between chemical and dynamical processes in the low and middle atmosphere under the frame of a CCM, a General Circulation Model (GCM), from the Institute of Numerical Mathematics of the Russian Academy of Sciences (INM RAS), is interactively coupled with a global chemistry-transport model (CTM) of the Russian State Hydrometeorological University (RSHU), which was designed based on the three-dimensional extension of the SUNY-SPB two-dimensional model [29–32]. The CCM calculates the variability of 74 chemically active gases in the lower and middle atmospheres, which are interacting in 174 gas-phase and heterogeneous chemical reactions and 51 photodissociation processes, including oxygen, nitrogen, hydrogen, chlorine, bromine, carbon, and sulfur cycles [33,34]. The model considers the processes of the formation and evolution of polar stratospheric clouds (PSC) based on stratospheric sulfate aerosol [35–37].

The influence of ozone-depleting substances on atmospheric chemistry is taken into account in the model by setting the concentrations of freons and halons at the lower boundary with subsequent calculation of their chemical evolution and photodissociation in the troposphere and stratosphere, taking into account their transport by advection and turbulent fluxes. The influence of spectral solar radiation fluxes is taken into account by specifying their variability at the upper boundary of the model (about 90 km), taking into account the attenuation of the overlying layers of the atmosphere and calculating their attenuation due to absorption and scattering by gases and aerosols. The total flux of direct and scattered radiation is further used to calculate both the rates of photodissociation of atmospheric gases and the heating of the atmosphere by short-wave radiation. The long-wave outgoing radiation of the earth's surface and atmosphere is calculated to estimate the cooling of the atmosphere. The aerosol influence was taken into account by specifying the aerosol surface area in a unit volume, which was used to calculate the rates of heterogeneous reactions, to estimate the optical thickness of the aerosol attenuation of solar radiation and as a basis for the formation of polar stratospheric clouds at low temperature. The sea surface temperature and the sea ice coverage are taken into account when setting the lower boundary condition for the thermodynamics equation in GCM, which affects the temperature and, accordingly, the rate of chemical reactions, as well as the wind speed and the transport of ozone and related gases.

To assess the role of the main factors influencing the interannual variability of ozone content, emissions of chlorine and bromine substances contributing to ozone depletion (ODS), set on the basis of the World Meteorological Organization (WMO) scenarios of 2011 [2]; changes in solar radiation fluxes at the upper atmosphere set according to the empirical Naval Research Laboratory (NRL) model [38]; stratospheric sulfate aerosol content, set according to [39], and sea surface temperature

(SST) set according to the Met-Office re-analysis [40], in addition to the baseline scenario, which takes into account the influence of all these factors on the interannual variability of ozone content (basic scenario). Additional model numerical experiments were carried out, which took into account the influence of only each of these factors separately. A comparison of the results of model experiments under these scenarios allowed us to estimate the role of each of the tested factors in the interannual variability of the total ozone content in different latitudinal zones. The interannual variability of the studied factors influencing ozone trends is shown in Figure 1.

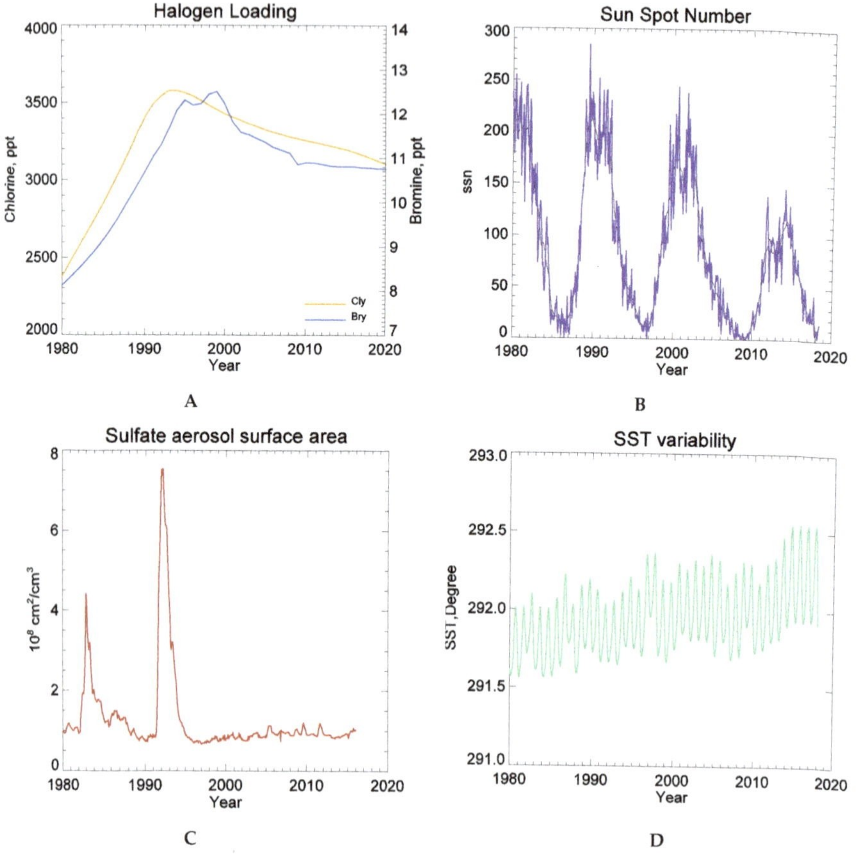

Figure 1. Interannual variability of the tested factors influencing ozone trends: (**A**) chlorine and bromine ozone destroyers; (**B**) sunspot number (SSN) variability [41]; (**C**) stratospheric aerosol; (**D**) sea surface temperature.

3. Results

3.1. Total Column Ozone Interannual Variability in the Tropics

Both satellite measurements and numerical simulations allow us to select three periods of interannual variability in total ozone in the tropics (Figure 2): a significant drop from the early 80s to the mid-90s of the twentieth century, a sharp increase in ozone from the mid-90s to the early 2000s, and weak variability during the first 20 years of the twenty-first century. This variability is superimposed by short-term fluctuations in total ozone with highs around 1990, 2000, and 2010, i.e., almost every 10 years, and lows around 1986–1987, 1993–1994, 2005–2007, and 2016. Trend estimates from SBUV satellite measurements for the entire study period (1980–2018) and separately in the twentieth and twenty-first centuries show a general trend of a weak decrease in total ozone over all three periods. At the same time, according to Fisher's criterion, these trends have a significance with a probability density function (PDF) of 92% for the general trend and the trend of the late twentieth century and 0.59% for the trend of the twenty-first century. The coefficients of the linear trend are as follows: period 1980–2018 −0.06 Dobson units (DU) per year, period 1980–2000 −0.17 DU per year, period 2000–2018 −0.07 DU per year. Thus, the trends for the entire study period and the twentieth century are significant with a 10% level of significance, and the trend of the twenty-first century can be considered insignificant. If we use our trend estimates from the three aforementioned periods, we find that the linear trend coefficient for 1980–1995 is −0.35 DU per year with PDF = 0.98 (i.e., a 5% significance level), and the coefficient for 1995–2000, when there was a sharp increase in the total ozone content, was 0.87 DU per year.

Analysis of the results of model experiments on the sensitivity of the tropical ozone to the variability of influencing factors (Figure 2B) shows that the basic tendency of ozone variability during the entire tested period is influenced by the increase in chlorine and bromine gases during the late twentieth century as well as their reduction in the early twenty-first century, determined on the basis of WMO 2011 scenarios. Comparisons with satellite measurements shown in Figure 2A reveal that WMO 2011 scenarios are likely overestimating the reduction of chlorine and bromine gases in the twenty-first century, which is reflected in a significant excess of the model results from satellite observations at the end of the tested period, namely in 2010–2020. At the same time, at the beginning of the period, i.e., 1980–1985, the importance of chlorine and bromine ozone depletion is probably also underestimated in WMO 2011 scenarios, which is also evidenced by a significant excess of model values from satellite observations. The period of rapid reduction of column ozone in the first half of the 90s, indicated both by SBUV observations and model calculations (Figure 2A), is associated not only with increased halogen loading with maximum effect on ozone at the end of the 90s (Figure 2B) but also with the influence of other factors, in particular, the increase in the content of sulfate stratospheric aerosol as a result of emissions by mount Pinatubo (peak aerosol effect on ozone at 1992) (Figure 2B), solar activity (ozone minimum at 1995–1996, Figure 2B), and sea surface temperature (peak influence on ozone at 1995). At the end of the 90s and beginning of the 2000s, when halogen loading was at a maximum (Figure 2B—orange line), both satellite data and model calculations revealed the increase in column ozone (Figure 2A).

Comparison of the Figure 1B with Figure 2A,B allows us to conclude that solar activity, particularly its 11-year cycle, is responsible for the observed positive peaks of tropical ozone in 1990, 2002–2003, and 2014–2015 as well as for the low column ozone in 1986–1987 and 2007–2008. The change in stratospheric sulfate aerosol loading (Figure 1C) plays an important role in ozone depletion not only after the Mt. Pinatubo emissions in the first part of the 90s but also after the emissions of the El Chichon in the mid-80s (Figure 2B). The sea surface temperature's impact on the column ozone is visible at the end of the 90s after the major El Nino southern oscillation (ENSO) event in 1997–1998 resulted in the enhanced SST (Figure 1D) and column ozone increase in the tropics (Figure 2B—green line). The changes caused by ENSO in the dynamics of ozone and related gas transport contribute to a sharp increase of tropical ozone in the late 90s of the twentieth century, along with solar activity and purification of the stratosphere from the volcanic aerosol. Overall, it should be noted that the results of the numerical experiments demonstrated that the changes in chlorine and bromine loading regulated by the Montreal Protocol and its amendments are responsible for long-term decadal variability in the column ozone, while its short-term changes with time scale around several years are mostly defined by natural factors, such as the 11-year solar cycle, stratosphere sulfate aerosol loading, and variability of the dynamic processes initiated by sea surface temperature changes. The simultaneous influence of these factors on the decreasing in the column ozone in the early 90s and increasing the column ozone in the latter part of the 90s led to drastic, sharp changes in the content of tropical ozone. At the same time, the influence of the anthropogenic chlorine–bromine factors in these processes is not dominant.

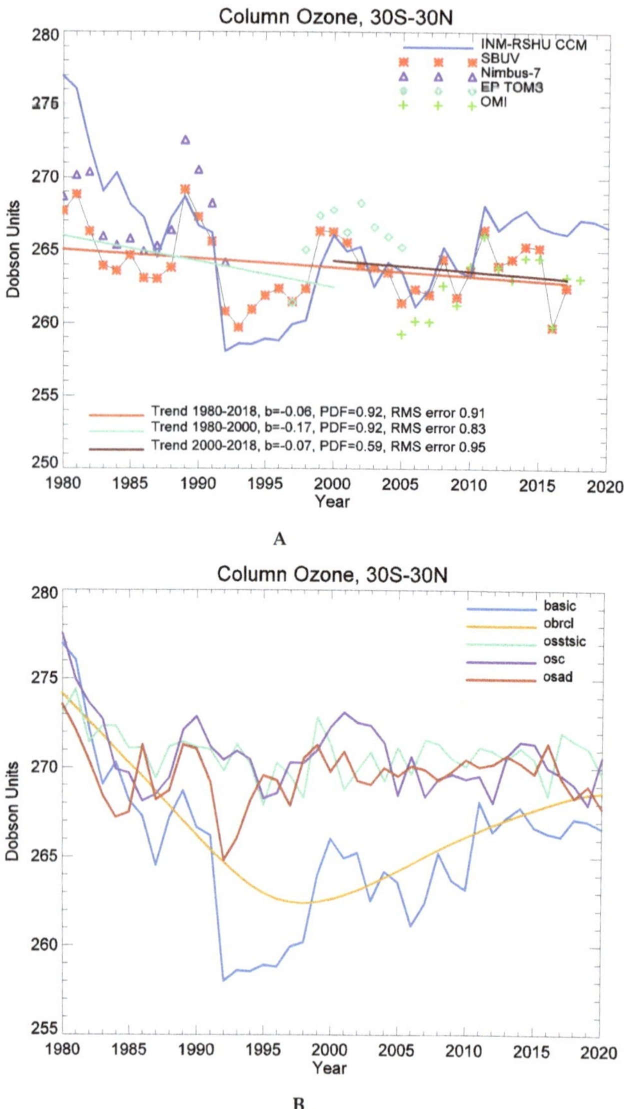

Figure 2. Interannual variability of column ozone in the tropics: (**A**) Results of satellite measurements (solar backscattering ultraviolet (SBUV), Nimbus-7 Total Ozone Mapping Spectrometer (TOMS), Earth Probe (EP) TOMS, and the Ozone Monitoring Instrument (OMI)) and numerical modeling (chemistry–climate model of the Institute of Numerical Mathematics and the Russian State Hydrometeorological University) with an assessment of trends (b—coefficient of linear regression) and their significance (PDF—probability density function, and RMS—root mean square error); (**B**) results of numerical modeling under scenarios with a separate action of influencing factors: basic—all factors' interannual variabilities are taken into account, obrcl—only the surface emissions of bromine and chlorine-containing gases' interannual variability is taken into account, osstsic—only sea surface temperature and sea ice coverage interannual variabilities are taken into account, osc—only solar radiation's interannual variability is taken into account, osad—only sulfate aerosol density's interannual variability is taken into account.

3.2. Total Column Ozone Interannual Variability in the Mid-Latitudes of the Northern Hemisphere

For the middle latitudes of the northern hemisphere (Figure 3), satellite observations indicate the same three periods of steady change in total column ozone as for tropical latitudes. Of particular note is a significant decrease in ozone from the early 1980s to the mid-1990s, a sharp increase in its content in the second half of the 1990s, and a weak declining in the column ozone in the twenty-first century. Quantitative trend estimates show a greater rate of decrease in the northern mid-latitude ozone compared to that of tropical latitudes. In particular, for the entire considered period of 1980–2020, the linear trend coefficient is −0.19 Dobson Units (DU) per year with PDF = 0.97. For the end of the twentieth century, the linear trend coefficient is −0.76 DU per year with PDF = −0.99, and for the beginning of the twenty-first century, the linear trend coefficient is −0.11 DU per year with PDF = 0.43. Thus, it can be stated that negative trends in the northern mid-latitude column ozone are significant with a level of 5% for the entire period of 1980–2020 and for the end of the twentieth century (1980–2000). For the beginning of the twenty-first century, the trend can be considered insignificant. Column ozone trend estimation for the three earlier allocated periods indicate that in 1980–1995, the coefficient of the linear trend is −1.41 DU per year with PDF = 0.99, and in 1995–2000, the coefficient of the positive linear trend is equal to 1.42 DU per year.

The analysis of factors affecting northern mid-latitude column ozone demonstrates (Figure 3B) that chlorine and bromine loading in accordance with WMO scenarios leads to significant column ozone reduction in the northern mid-latitudes during the first half of the research period (1980–1995) and to column ozone rise at the end of the research period (2010–2020). From the mid-90s to 2010, the influence of halogen gases on the northern mid-latitude column ozone is negligible. Other factors include the greater influence of stratospheric aerosol at the end of the twentieth century compared to tropical latitudes, especially during periods of volcanic eruptions, and the greater role of sea surface temperature variability comparable to solar activity. In periods of sharp decrease and then ozone recovery in the 90s similar to tropical latitudes, a combination of several factors played an additive role: the fall in ozone as a result of increased chlorine and bromine gas content, the role of sea surface temperature on the background of increasing chlorine and bromine gases, the increase in the ozone content as a result of the purification of stratospheric aerosol, the influence of the variability of sea surface temperature, and the rise and reduction of solar activity.

3.3. Total Column Ozone Interannual Variability in the Mid-Latitudes of the Southern Hemisphere

The column ozone variability in the southern middle latitudes (Figure 4) is, on the one hand, characterized by the presence of a fairly pronounced negative trend throughout the period of 1980–2020 with a higher coefficient of −0.33 DU per year and PDF = 0.99. On the other hand, the periods of sharp decrease and then increase in ozone content in the 90s, clearly distinguished in the tropics and mid-latitudes of the northern hemisphere and in the Southern hemisphere, have a different look according to the SBUV (Figure 4A). In particular, there are two column ozone minima in 1993 and 1997, two maxima in 1996 and 1998, and growth in the late twentieth century that continues until 2002. As a result, if, as was done for the tropical and middle latitudes of the Northern hemisphere, we were to evaluate trends from 1980 to 1995 and then from 1995 to 2000, the trend in 1980–1995 differs little from the trend for the overall period of 1980–2000 (coefficient −1.05 DU per year vs. −0.9 DU per year), and there is an insignificant column ozone trend (PDF = 0.58) from 1995 to 2000.

Thus, for the middle latitudes of the southern hemisphere, it is possible to propose a different allocation of periods of steady change in ozone content compared to the other latitudes: the period of decrease in ozone content is extended from 1980 to 1997, and the period of growth covers the years from 1997 to 2002, and third period from 2002 to 2018. Then, for these periods, the linear trend coefficients are −1.0, 2.0, and −0.01 DU per year, respectively, and the values of PDF are 0.99, 0.97, and 0.05, respectively. In addition, for the twenty-first century in the mid-latitudes of the Southern hemisphere, a quasi-two-year cycle of changes in the total ozone content is noticeable.

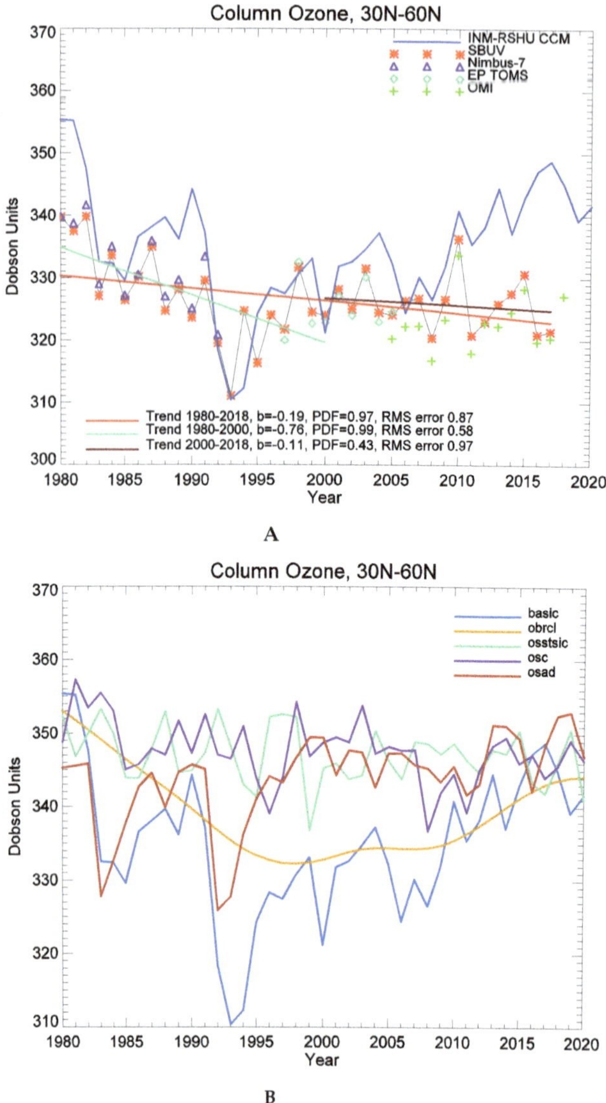

Figure 3. Interannual variability of column ozone in the northern mid-latitudes: (**A**) Results of satellite measurements (solar backscattering ultraviolet (SBUV), Nimbus-7 Total Ozone Mapping Spectrometer (TOMS), Earth Probe (EP) TOMS, and the Ozone Monitoring Instrument (OMI)) and numerical modeling (chemistry—climate model of the Institute of Numerical Mathematics and the Russian State Hydrometeorological University) with an assessment of trends (b—coefficient of linear regression) and their significance (PDF—probability density function, and RMS—root mean square error); (**B**) results of numerical modeling under scenarios with a separate action of influencing factors: basic—all factors' interannual variabilities are taken into account, obrcl—only the surface emissions of bromine and chlorine-containing gases' interannual variability is taken into account, osstsic—only sea surface temperature and sea ice coverage interannual variabilities are taken into account, osc—only solar radiation's interannual variability is taken into account, osad—only sulfate aerosol density's interannual variability is taken into account.

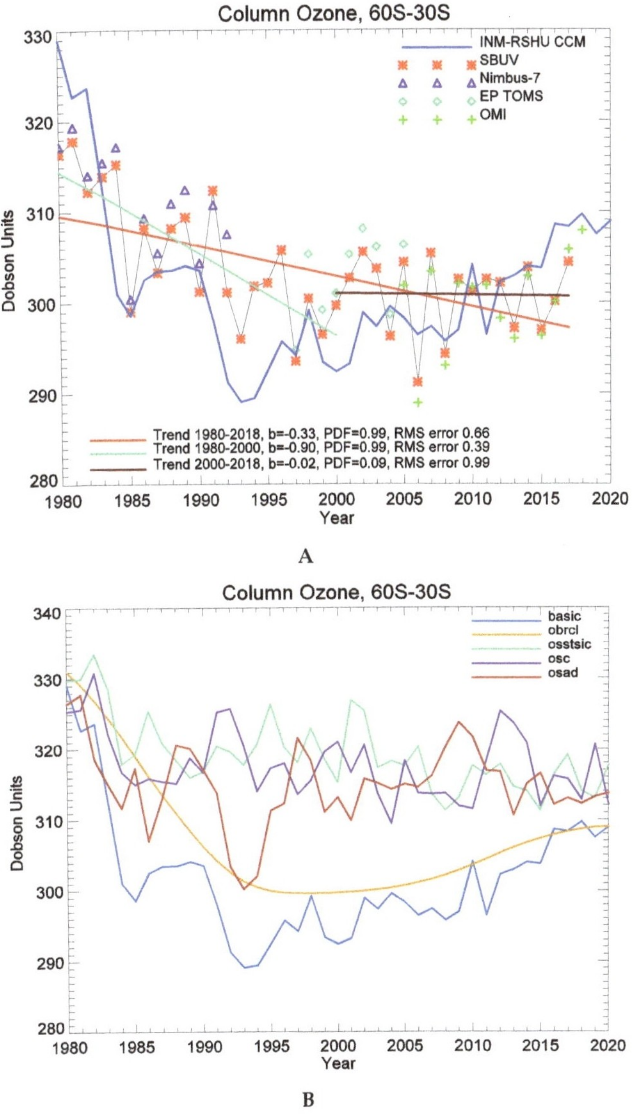

Figure 4. Interannual variability of column ozone in the southern mid-latitudes: (**A**) Results of satellite measurements (solar backscattering ultraviolet (SBUV), Nimbus-7 Total Ozone Mapping Spectrometer (TOMS), Earth Probe (EP) TOMS, and the Ozone Monitoring Instrument (OMI)) and numerical modeling (chemistry—climate model of the Institute of Numerical Mathematics and the Russian State Hydrometeorological University) with an assessment of trends (b—coefficient of linear regression) and their significance (PDF—probability density function, and RMS—root mean square error); (**B**) results of numerical modeling under scenarios with a separate action of influencing factors: basic—all factors' interannual variabilities are taken into account, obrcl—only the surface emissions of bromine and chlorine-containing gases' interannual variability is taken into account, osstsic—only sea surface temperature and sea ice coverage interannual variabilities are taken into account, osc—only solar radiation's interannual variability is taken into account, osad—only sulfate aerosol density's interannual variability is taken into account.

Analysis of influencing factors ((Figure 4B) reveals that stratospheric sulfate aerosol in the mid-latitudes of the southern hemisphere plays a smaller role than noted both in the tropics and in the northern hemisphere, with the model likely overestimating the importance of this factor in the southern hemisphere. The impact of chlorine and bromine loading according to WMO 2011 scenarios on changes in the southern hemisphere's mid-latitude ozone is also overestimated by the model, especially at the end of the period (2010–2020). Interestingly, the influence of solar activity is manifested differently in the mid-latitudes of the northern and southern hemispheres, despite the fact that the variability of the spectral fluxes of solar radiation is the same for the entire globe, according to NRL data. In particular, in a scenario that takes into account only the variability of solar flows (Figure 4B), there are significant ozone peaks in 1991, 1997, 2002, and 2005, which are also evident in satellite observations. These peaks are not obviously related to the 11-year solar cycle, but more to the short-term variability of the solar fluxes (Figure 1B). In the model scenario, taking into account the influence of all factors, some of these maxima are offset by the influence of other factors, in particular the overestimated role of stratospheric sulfate aerosol. The effect of sea surface temperature variability is comparable in order of magnitude to that of solar activity and, in some years, enhances it (e.g., 2002) while in some years it compensates for it (e.g., 2000). Overall, the model's overestimation of the chemical factors of ozone depletion in the southern mid-latitudes is, probably, a result of underestimating the southern dynamical processes' roles in the model as well as overestimating the Antarctic ozone chemical depletion, visible in Figure 6.

3.4. Total Column Ozone Interannual Variability in the Polar Regions

In the Arctic zone (Figure 5), the general trend of ozone content change is much smaller than observed at other latitudes (the linear trend coefficient for 1980–2020 is −0.22 DU per year) with relatively low significance (PDF = 0.83), whereas the twentieth century trend has a coefficient of −1.21 DU per year and PDF = 0.99, and the twenty-first century's trend is negative and insignificant (PDF = 0.18).

The analysis of influencing factors (Figure 5B) shows that, first, as in other regions in the Arctic, the role of chlorine and bromine gases is overestimated in 2010–2020, resulting in a significant excess of simulation results over satellite data. Secondly, the important role of dynamic factors in the model experiments reflect the influence of sea surface temperature on the column ozone, probably due to the impact of dynamical processes on the stability of the polar vortex and heat and mass exchange between the Arctic and mid-latitudes. Thirdly, there is a combined influence of solar activity and stratospheric aerosol, which in some years gives the strengthened maxima coinciding with phases of mid-latitudes, but is not visible in the results of satellite observations in the Arctic (as for example in 1990). This indicates the underestimated role of dynamic factors in the model.

In Antarctica (Figure 6), on the one hand, there are two periods: the first with a significant negative trend of column ozone from 1980 to 2000 (linear trend coefficient −2.26 DU per year and PDF = 0.99), and, on the other hand, an insignificant positive trend in the twenty-first century (PDF = 0.56). The results of the simulation, in contrast to most other regions in the Antarctic, are substantially underestimating the ozone content.

While assessing the influencing factors (Figure 6B), first of all, it should be noted that, unlike other regions, the impact of chlorine and bromine loading does not lead to significant column ozone rise at the end of the considered period (2010–2020), whereas at the beginning of this period (1980–1995), the influence of halogen gases is maximum compared to other latitudes (compare the orange lines in Figures 2B, 3B, 4B, 5B and 6B). Thus, the weak sign of Antarctic ozone recovery in the twenty-first century may not merely be related to measures directed to reducing the emissions of ozone-depleting substances.

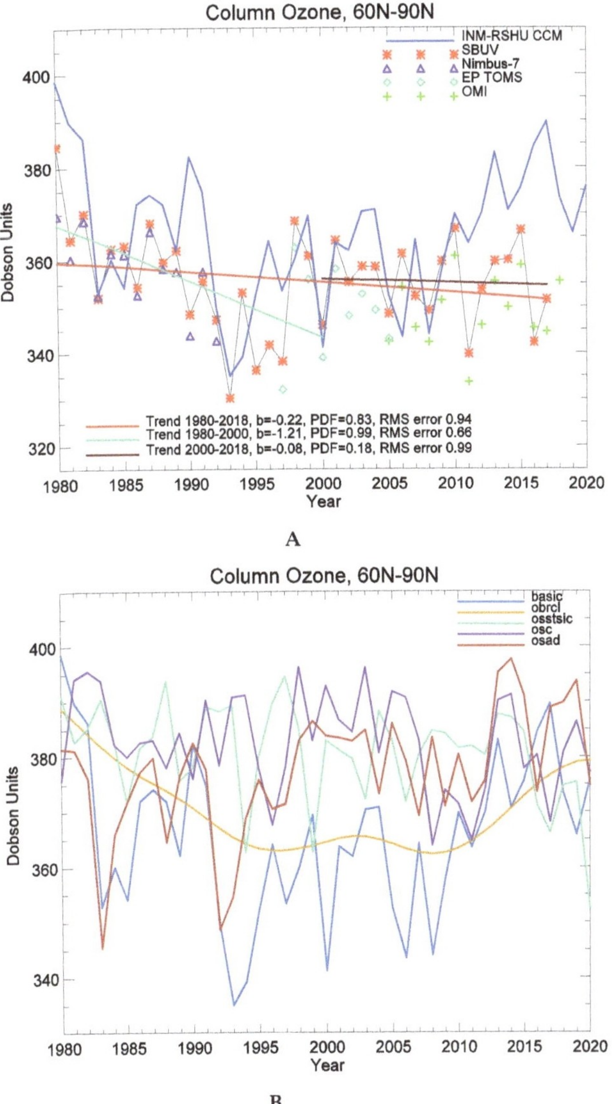

Figure 5. Interannual variability of column ozone in the northern polar latitudes: (**A**) Results of satellite measurements (solar backscattering ultraviolet (SBUV), Nimbus-7 Total Ozone Mapping Spectrometer (TOMS), Earth Probe (EP) TOMS, and the Ozone Monitoring Instrument (OMI)) and numerical modeling (chemistry–climate model of the Institute of Numerical Mathematics and the Russian State Hydrometeorological University) with an assessment of trends (b—coefficient of linear regression) and their significance (PDF—probability density function, and RMS—root mean square error); (**B**) results of numerical modeling under scenarios with a separate action of influencing factors: basic—all factors' interannual variabilities are taken into account, obrcl—only the surface emissions of bromine and chlorine-containing gases' interannual variability is taken into account, osstsic—only sea surface temperature and sea ice coverage interannual variabilities are taken into account, osc—only solar radiation's interannual variability is taken into account, osad—only sulfate aerosol density's interannual variability is taken into account.

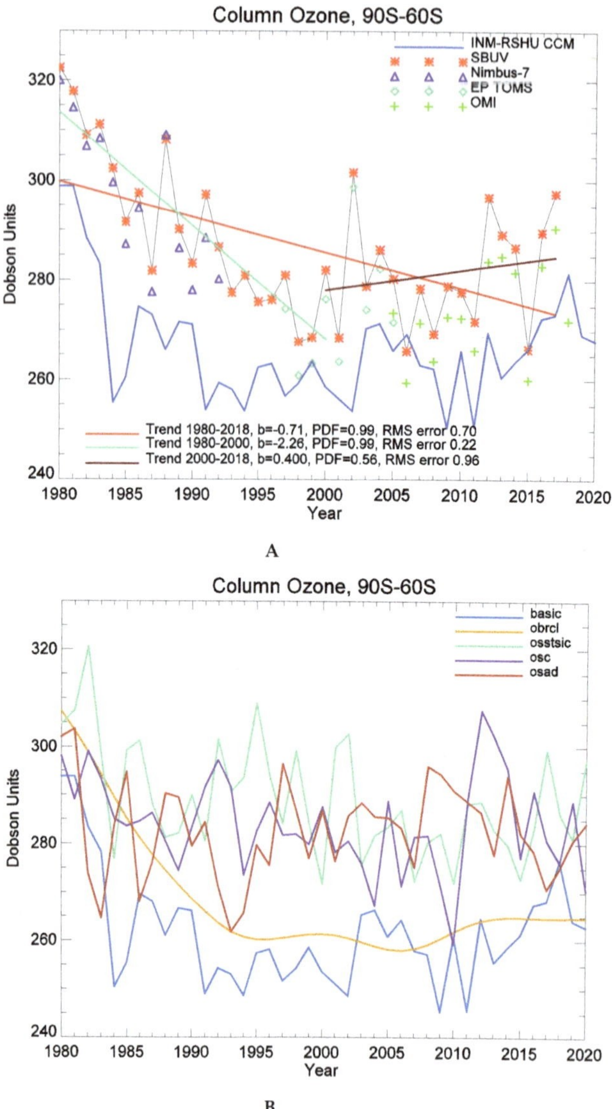

Figure 6. Interannual variability of column ozone in the southern polar latitudes: (**A**) Results of satellite measurements (solar backscattering ultraviolet (SBUV), Nimbus-7 Total Ozone Mapping Spectrometer (TOMS), Earth Probe (EP) TOMS, and the Ozone Monitoring Instrument (OMI)) and numerical modeling (chemistry–climate model of the Institute of Numerical Mathematics and the Russian State Hydrometeorological University) with an assessment of trends (b—coefficient of linear regression) and their significance (PDF—probability density function, and RMS—root mean square error); (**B**) results of numerical modeling under scenarios with a separate action of influencing factors: basic—all factors' interannual variabilities are taken into account, obrcl—only the surface emissions of bromine and chlorine-containing gases' interannual variability is taken into account, osstsic—only sea surface temperature and sea ice coverage interannual variabilities are taken into account, osc—only solar radiation's interannual variability is taken into account, osad—only sulfate aerosol density's interannual variability is taken into account.

4. Conclusions and Discussion

In this paper, the study of the column ozone in the tropical, mid, and polar latitudes is presented for the period from 1980 to 2018. Analyses of the results of satellite observations demonstrated that for all considered latitudinal regions, there is a significant negative trend in the annual average column ozone during the entire study period. At the same time, the level of significance for the tropics is 10%, for the middle latitudes of the northern and southern hemispheres and Antarctica is 5%, and for the Arctic zone it is 20%. Thus, the least significant negative trend is observed in the northern polar zone, and the most significant negative trend of total ozone (PDF = 0.99) is in the middle and polar latitudes of the southern hemisphere.

When considering individual trends for the variability of total ozone in the twentieth and twenty-first centuries of the considered time period for all latitudinal zones, a significant negative trend is found at the end of the twentieth century (1980–2000) and an insignificant trend is at the beginning of the twenty-first century (2000–2018). At the same time, a positive albeit insignificant trend is observed only in Antarctica in the twenty-first century. The significance level for all latitudinal zones except tropical for the late twentieth century is 1%, while in the tropical zone it is 10%. A more detailed examination of interannual variability allows us to distinguish in all three latitudinal bands a period of steady variability in the column ozone: a strong decrease since the beginning of the 80s until the mid-90s, the rapid growth since the mid-90s to early twenty-first century, and then an insignificant column ozone trend in the twenty-first century. At the same time, while in the northern hemisphere and the tropics the duration of the first period of rapid ozone depletion is 13–15 years (until 1993–1995), in the southern hemisphere this period lasts until 1997–1998, i.e., 3–4 years longer. In addition, the period of rapid growth in the late 90s to the early 2000s is also different for the two hemispheres. In the northern hemisphere, it ends almost at the turn of the century, and it lasts two years longer in the southern hemisphere (until 2002).

An analysis of the significance of individual factors, based on the results of numerical experiments with the chemical-climate model, showed that WMO 2011 scenarios probably overestimating the reduction of ozone-depleting gas emissions into the atmosphere in 2010–2020. As a result, model calculations for all latitudinal zones depict a fairly rapid increase in column ozone, while satellite observations do not confirm this. In the tropics, the rise in column ozone due to reduction in the halogen loading from the late 90s till 2020 is estimated as uniform with a total of 7 Dobson units. In contrast, in the middle latitudes of the northern hemisphere until 2010, the model reveals a weak increase of 2 Dobson units and then a stronger increase of 10 Dobson units from 2010 to 2020. In the mid-latitudes of the southern hemisphere, the halogen factor has little effect on ozone from 1995 to 2006, and then leads to an increase in ozone at 9 Dobson units by 2020. In both polar regions from 1995 to 2010, the ozone content changes very little despite changes in emissions of chlorine and bromine gases. Finally, there is a 16 Dobson-unit rise in column ozone in the Arctic, and a 3 Dobson-unit rise in Antarctica.

Other factors include the predominance of solar radiation flux variability in the tropics and the mid-latitudes of the southern hemisphere, stratospheric aerosol loading in the mid-latitudes of the northern hemisphere, and sea surface temperature in the polar regions. The influence of quasi-biennial oscillation in the mid-latitudes of the southern hemisphere in the twenty-first century should also be noted, and are quite clearly presented in the results of satellite observations and only slightly noticeable in the results of modeling.

Author Contributions: Conceptualization, S.P.S., V.Y.G.; Methodology, S.P.S., V.Y.G.; Formal analysis, P.A.B.; Data curation, A.R.J.; Writing—original draft preparation, S.P.S.; Writing—review and editing, S.P.S.; Visualization, A.R.J. All authors have read and agreed to the published version of the manuscript.

Funding: This research was funded by the Russian Science Foundation, grant: 19-17-00198. Polar processes were studied under the Russian Foundation for Basic Research, grant: 17-05-01277.

Acknowledgments: We would like to thank National Aeronautics and Space Administration's Goddard Space Flight Center for providing the SBUV Merged Ozone Data Set, and Atmospheric Chemistry and Dynamics

Laboratory (Code 614) for providing us with the Nimbus-7 and Earth Probe TOMS data as well as the Aura OMI instrument column total ozone data.

Conflicts of Interest: The authors declare no conflict of interest.

References

1. WMO (World Meteorological Organization). *Scientific Assessment of Ozone Depletion: 2006, Global Ozone Research and Monitoring Project—Report No. 50*; WMO: Geneva, Switzerland, 2007; 572p.
2. WMO (World Meteorological Organization). *Scientific Assessment of Ozone Depletion: 2010, Global Ozone Research and Monitoring Project—Report No. 52*; WMO: Geneva, Switzerland, 2011; 516p.
3. WMO. *Scientific Assessment of Ozone Depletion: 2014 Global Ozone Research and Monitoring Project Report*; World Meteorological Organization: Geneva, Switzerland, 2014; p. 416.
4. Solomon, S. Stratospheric ozone depletion: A review of concepts and history. *Rev. Geophys.* **1999**, *37*, 275–316. [CrossRef]
5. Andersen, S.B.; Weatherhead, E.C.; Stevermer, A.; Austin, J.; Brühl, C.; Fleming, E.L.; De Grandpré, J.; Grewe, V.; Isaksen, I.; Pitari, G.; et al. Comparison of recent modeled and observed trends in total column ozone. *J. Geophys. Res.* **2006**, *111*, 4428. [CrossRef]
6. Molina, M.J.; Rowland, F.S. Stratospheric sink for chlorofluoromethanes: Chlorine atomc-atalysed destruction of ozone. *Nature* **1974**, *249*, 810–812. [CrossRef]
7. Solomon, P.; Barrett, J.; Mooney, T.; Connor, B.; Parrish, A.; Siskind, D.E. Rise and decline of active chlorine in the stratosphere. *Geophys. Res. Lett.* **2006**, *33*, L18807. [CrossRef]
8. Chipperfield, M.P.; Bekki, S.; Dhomse, S.; Harris, N.R.; Hassler, B.; Hossaini, R.; Steinbrecht, W.; Thiéblemont, R.; Weber, M. Detecting recovery of the stratospheric ozone layer. *Nature* **2017**, *549*, 211–218. [CrossRef]
9. Frith, S.M.; Kramarova, N.A.; Stolarski, R.S.; McPeters, R.D.; Bhartia, P.K.; Labow, G.J. Recent changes in total column ozone based on the SBUV Version 8.6 Merged Ozone Data Set. *J. Geophys. Res. Atmos.* **2014**, *119*, 9735–9751. [CrossRef]
10. Harris, N.R.P.; Hassler, B.; Tummon, F.; Bodeker, G.E.; Hubert, D.; Petropavlovskikh, I.; Steinbrecht, W.; Anderson, J.; Bhartia, P.K.; Boone, C.D.; et al. Past changes in the vertical distribution of ozone—Part 3: Analysis and interpretation of trends. *Atmos. Chem. Phys.* **2015**, *15*, 9965–9982. [CrossRef]
11. Weber, M.; Coldewey-Egbers, M.; Fioletov, V.E.; Frith, S.M.; Wild, J.D.; Burrows, J.P.; Long, C.S.; Loyola, D. Total ozone trends from 1979 to 2016 derived from five merged observational datasets—The emergence into ozone recovery. *Atmos. Chem. Phys.* **2018**, *18*, 2097–2117. [CrossRef]
12. Zvyagintsev, A.M.; Vargin, P.N.; Peshin, S. Total ozone variations and trends during the period 1979–2014. *Atmos. Ocean. Opt.* **2015**, *28*, 575–584. [CrossRef]
13. Sofieva, V.F.; Kyrölä, E.; Laine, M.; Tamminen, J.; Degenstein, D.; Bourassa, A.; Roth, C.; Zawada, D.; Weber, M.; Rozanov, A.; et al. Merged SAGE II, Ozone_cci and OMPS ozone profile dataset and evaluation of ozone trends in the stratosphere. *Atmos. Chem. Phys.* **2017**, *17*, 12533–12552. [CrossRef]
14. Chehade, W.; Weber MBurrows, J.P. Total ozone trends and variability during 1979–2012 from merged data sets of various satellites. *Atmos. Chem. Phys.* **2014**, *14*, 7059–7074. [CrossRef]
15. Ball, W.T.; Alsing, A.; Mortlock, D.J.; Staehelin, J.; Haigh, J.D.; Peter, T.; Tummon, F.; Stübi, R.; Stenke, A.; Anderson, J.; et al. Continuous decline in lower stratospheric ozone offsets ozone layer recovery. *Atmos. Chem. Phys.* **2018**, *18*, 1379–1394. [CrossRef]
16. Zubov, V.; Rozanov, E.; Egorova, T.; Karol, I.; Schmutz, W. Role of external factors in the evolution of the ozone layer and stratospheric circulation in 21st century. *Atmos. Chem. Phys.* **2013**, *13*, 4697–4706. [CrossRef]
17. Geller, A.M.; Smyshlyaev, S.P. A model study of total ozone evolution 1979–2000—The role of individual natural and anthropogenic effects. *Geophys. Res. Lett.* **2002**, *29*, 2048. [CrossRef]
18. Robinson, S.A.; Wilson, S.R. *Environmental Effects of Ozone Depletion and Its Interactions with Climate Change: 2010 Assessment*; United Nations Environment Programme: Nairobi, Kenya, 2010; 328p.
19. Solomon, S.; Ivy, D.J.; Kinnison, D.; Mills, M.J.; Neely, R.R.; Schmidt, A. Emergence of healing in the Antarctic ozone layer. *Science* **2016**, *353*, 269–274. [CrossRef] [PubMed]
20. Heath, D.F.; Krueger, A.J.; Roeder, H.A.; Henderson, B.D. Solar backscatter ultraviolet and total ozone mapping spectrometer (SBUV/TOMS) for Nimbus G. *Opt. Eng.* **1975**, *14*, 323–331. [CrossRef]

21. Bhartia, P.K.; McPeters, R.D.; Flynn, L.E.; Taylor, S.; Kramarova, N.A.; Frith, S.; Fisher, B.; DeLand, M. Solar Backscatter UV (SBUV) total ozone and profile algorithm. *Atmos. Meas. Tech.* **2013**, *6*, 2533–2548. [CrossRef]
22. McPeters, R.D.; Bhartia, P.K.; Krueger, A.J.; Herman, J.R.; Schlesinger, B.M.; Wellemeyer, C.G.; Seftor, C.J.; Jaross, G.; Taylor, S.L.; Swissler, T.; et al. *Nimbus-7 Total Ozone Mapping Spectrometer (TOMS) Data Product's User's Guide*; National Aeronautics and Space Administration: Washington, DC, USA, 1996.
23. McPeters, R.D.; Hollandsworth, S.M.; Flynn, L.E.; Herman, J.R.; Seftor, C.J. Long-Term Ozone Trends Derived From the 16-Year Combined Nimbus7/Meteor 3 TOMS Version 7 Record. *Geophys. Res. Lett.* **1996**, *23*, 3699–3702. [CrossRef]
24. McPeters, R.D.; Labow, G.J. An Assessment of the Accuracy of 14.5 Years of Nimbus 7 TOMS Version 7 Ozone Data by Comparison with the Dobson Network. *Geophys. Res. Lett.* **1996**, *23*, 3695–3698. [CrossRef]
25. McPeters, R.D.; Bhartia, P.K.; Krueger, A.J.; Herman, J.R. *Earth Probe Total Ozone Mapping Spectrometer (TOMS) Data Products User's Guide*; National Aeronautics and Space Administration: Washington, DC, USA, 1998.
26. Levelt, P.F.; Hilsenrath, E.; Leppelmeier, G.W.; van den Oord, G.H.J.; Bhartia, P.K.; Tamminen, J.; de Haan, J.F.; Veefkind, J.P. Science objectives of the ozone monitoring instrument. *Geosci. Remote Sens.* **2006**, *44*, 1199–1208. [CrossRef]
27. Wilks, D.S. *Statistical Methods in the Atmospheric Sciences*; International Geophysics Series; Academic Press Elsevier Inc.: Oxford, UK, 2011; 676p.
28. Galin, V.Y.; Smyshlyaev, S.P.; Volodin, E.M. Combined chemistry-climate model of the atmosphere. *Izv. Atmos. Ocean. Phys.* **2007**, *43*, 399–412. [CrossRef]
29. Smyshlyaev, S.P.; Dvortsov, V.L.; Geller, M.A.; Yudin, V. A two-dimensional model with input parameters from a GCM: Ozone sensitivity to different formulations for the longitudinal temperature variation. *J. Geophys. Res.* **1998**, *103*, 28373–28387. [CrossRef]
30. Dvortsov, V.L.; Geller, M.A.; Yudin, V.; Smyshlyaev, S. Parameterization of the convective transport in a 2-D chemistry-transport model and its validation with Radon 222 and other tracer simulations. *J. Geophys. Res.* **1998**, *103*, 22047–22062. [CrossRef]
31. Smyshlyaev, S.P.; Geller, M.A.; Yudin, V.A. Sensitivity of model assessments of HSCT effects on stratospheric ozone resulting from uncertaintes in the NO_X production from lightning. *J. Geophys. Res.* **1999**, *104*, 401–418. [CrossRef]
32. Yudin, V.A.; Smyshlyaev, S.P.; Geller, M.A.; Dvortsov, V. Transport diagnostics of GCMs and implications for 2-D chemistry-transport model of troposphere and stratosphere. *J. Atmos. Sci.* **2000**, *57*, 673–699. [CrossRef]
33. Smyshlyaev, S.P.; Geller, M.A. Analysis of SAGE II observations using data assimilation by SUNY-SPB two-dimensional model and comparison to TOMS data. *J. Geophys. Res.* **2001**, *106*, 327–335. [CrossRef]
34. Diansky, R.A.; Galin, V.Y.; Gusev, A.V.; Smyshlyaev, S.P.; Volodin, E.M. The model of the Earth system developed at the INM RAS. *Russ. J. Numer. Anal. Math. Model.* **2010**, *25*, 419–429. [CrossRef]
35. Smyshlyaev, S.P.; Galin, V.Y.; Shaariibuu, G.; Motsakov, M.A. Modeling the Variability of Gas and Aerosol Components in the Stratosphere of Polar Regions. *Izv. Atmos. Ocean. Phys.* **2010**, *46*, 265–280. [CrossRef]
36. De Zafra, R.; Smyshlyaev, S. On the formation of HNO_3 in the Antarctic mid-to-upper stratosphere in winter. *J. Geophys. Res.* **2001**, *106*, 23115–23125. [CrossRef]
37. Sovde, A.; Gauss, M.; Smyshlyaev, S.; Isaksen, I.S.A. The Oslo CTM2: A Global Chemical Transport Model with Tropospheric and Stratospheric Chemistry. *J. Geophys. Res.* **2008**, *113*, 304. [CrossRef]
38. Dewolfe, W.A.; Wilson, A.; Lindholm, D.M.; Pankratz, C.K.; Snow, M.A.; Woods, T.N. Solar Irradiance Data Products at the LASP Interactive Solar IRradiance Datacenter (LISIRD). *AGU Fall Meet. Abstr.* **2010**, *21*, GC21B–0881.
39. Thomason, L.W.; Earnest, N.; Millán, L.; Rieger, L.; Bourassa, A.; Vernier, J.P.; Manney, G.; Luo, B.; Arfeuille, F.; Peter, T. A global space-based stratospheric aerosol climatology: 1979–2016. *Earth Syst. Sci. Data* **2018**, *10*, 469–492. [CrossRef]

40. Rayner, N.A.; Parker, D.E.; Horton, E.B.; Folland, C.K.; Alexander, L.V.; Rowell, D.P.; Kent, E.C.; Kaplan, A. Global analyses of sea surface temperature, sea ice, and night marine air temperature since the late nineteenth century. *J. Geophys. Res. Atmos.* **2003**, *108*, 4407. [CrossRef]
41. Sunspot Index and Long-Term Solar Observations (SILSO) Data/Image. Royal Observatory of Belgium: Belgium, Brussels. Available online: http://www.sidc.be/silso/home (accessed on 15 December 2019).

Article

Ozone Layer Evolution in the Early 20th Century

Tatiana Egorova [1], Eugene Rozanov [1,2,*], Pavle Arsenovic [3] and Timofei Sukhodolov [1]

1 PMOD/WRC, 7260 Davos, Switzerland; t.egorova@pmodwrc.ch (T.E.); timofei.sukhodolov@pmodwrc.ch (T.S.)
2 IAC ETH, 8092 Zurich, Switzerland
3 EMPA, 8600 Dübendorf, Switzerland; pavle.arsenovic@empa.ch
* Correspondence: eugene.rozanov@pmodwrc.ch

Received: 30 December 2019; Accepted: 3 February 2020; Published: 6 February 2020

Abstract: The ozone layer is well observed since the 1930s from the ground and, since the 1980s, by satellite-based instruments. The evolution of ozone in the past is important because of its dramatic influence on the biosphere and humans but has not been known for most of the time, except for some measurements of near-surface ozone since the end of the 19th century. This gap can be filled by either modeling or paleo reconstructions. Here, we address ozone layer evolution during the early 20th century. This period was very interesting due to a simultaneous increase in solar and anthropogenic activity, as well as an observed but not explained substantial global warming. For the study, we exploited the chemistry-climate model SOCOL-MPIOM driven by all known anthropogenic and natural forcing agents, as well as their combinations. We obtain a significant global scale increase in the total column ozone by up to 12 Dobson Units and an enhancement of about 20% of the near-surface ozone over the Northern Hemisphere. We conclude that the total column ozone changes during this period were mainly driven by enhanced solar ultra violet (UV) radiation, while near-surface ozone followed the evolution of anthropogenic ozone precursors. This finding can be used to constrain the solar forcing magnitude.

Keywords: ozone layer evolution; modeling; climate change; solar forcing; ozone precursors

---

1. Introduction

The state of the ozone layer has recently attracted growing attention in connection with the profound reduction of the global total ozone content in the 1980s, and an understanding of the role of the ozone layer not only as a defender of life on Earth from the damaging effects of hard solar ultraviolet radiation, but also as a factor affecting the climate and biosphere in general [1,2]. The evaluation of the ozone layer is well covered since the 1930s from the ground and, since the 1980s, by space-based instruments [3]. However, predicting future ozone behavior requires an understanding of its evolution in the past, when the combination of anthropogenic and natural factors affecting the ozone layer was different from the present day. Information about the state of the ozone layer in the past has not been available for most of the time, except for some measurements of near-surface ozone at the end of the 19th century [4–6]. This gap can be filled either by modeling or reconstructions from different proxies. The modeling efforts are mostly aimed at understanding the ozone changes between short periods during the preindustrial and present times. These changes were driven by a strong influence of manmade halogen containing ozone-depleting substances (hODS) on stratospheric ozone and enhanced anthropogenic emissions of tropospheric ozone precursors [7–14].

The continuous evolution of the ozone layer from the preindustrial to the present, including the first half of the 20th century, has been studied with numerical [15] and statistical [16] models. The global and annual mean total column ozone (TOC) increase during the 1900 to 1950 period simulated in [15] was about 0.2 Dobson Units (DU). This change was mostly driven by CO_2 induced cooling followed by slower ozone destruction cycles and the increase in tropospheric ozone due to elevated methane abundance. The obtained effects are not complete because the forcing was limited by the greenhouse gases, while the influence of enhanced solar activity [17] and tropospheric ozone precursors [10] were not considered. A statistical model that considers ozone-depleting substances, anthropogenic greenhouse gases, and natural processes that influence ozone was developed and used by [16] to obtain TOC evolution from 1900 to 2100. The model showed good performance in simulating observed TOC behavior. The use of solar activity as a proxy for the statistical model estimated a TOC increase of up to three DU during the early 20th century. However, the accuracy of this estimate was also limited by the absence of proxies related to energetic particles, the solar irradiance in different spectral bands, and tropospheric ozone precursors. Therefore, the ozone behavior in the past, before the emergence of manmade hODS, was not properly covered.

To fill these gaps in knowledge, we address the ozone layer evolution during the early 20th century, which is very interesting due to a simultaneous increase in solar and anthropogenic activity, the absence of powerful volcanic eruptions, as well as an observed, but not explained, substantial global warming [18,19]. This study is important for an understanding of the pattern of the ozone layer evolution in the past and to define which factors are the most important for the assessment of ozone layer evolution in the future. For this study, we exploited the chemistry-climate model (CCM) SOCOL-MPIOM driven by all known anthropogenic and natural forcing agents, as well as their combinations [19].

2. Experiments

The CCM SOCOL3-MPIOM [20–22] consists of the following three interactively coupled components: ECHAM5.4 [23] for the calculation of the atmospheric state, the chemistry module MEZON [24,25], and the ocean model MPIOM [26,27]. The CCM SOCOL3-MPIOM has T31 spectral horizontal resolution and covers the atmosphere from the ground to 0.01 hPa (~80 km).

We use the free running model version prescribing only the quasi-biannual oscillation in tropical zonal wind, which is not reproduced at the applied vertical resolution. The solar radiation forcing was prescribed according to a reconstruction [28] in six spectral intervals of our radiation code as follows: 180 to 250 nm, 250 to 440 nm, 440 to 660 nm, 660 to 1190 nm, 1190 to 2380 nm, and 2380 to 4000 nm. The reconstruction of total solar irradiance (TSI) in [28] gave a significant increase of ~1 W/m^2 per decade for the period from 1900 to 1950. This scenario gives a much larger solar irradiance forcing than the other available reconstructions [17] due to different assumption about the temporal variability of solar irradiance from the quiet Sun. The part of the solar heating rates missed in the ECHAM5 radiation code [29], and the photolysis rates, are calculated from the same solar irradiance reconstruction. Daily ionization rates by different precipitating energetic particles, as well as reactive nitrogen influx from the auroral regions, are prescribed according to recommendations for Coupled Model Intercomparison Project Phase 6 (CMIP6) [30]. The evolution of greenhouse gases, ozone-depleting substances, aerosol properties, and tropospheric ozone precursor emissions (CO and NO_x) are prescribed following [31]. The applied forcing is illustrated in [19].

With the CCM SOCOL3-MPIOM, we carried out seven ten-member ensemble model simulations covering the 1851 to 1940 period. The first experiment (referred hereafter as ALL) included all available observed and reconstructed forcing agents. To investigate the contributions of all considered forcings, we either fixed them at 1851 values or excluded them completely. For the second simulation, we eliminated the energetic particle precipitation (noEPP). The third experiment was driven by the same forcing as in ALL, but the solar irradiance in the 180 to 250 nm band, extra heating, and photolysis rates were fixed at 1851 values. This experiment, named fixUV (fixed solar ultraviolet), was designed to eliminate all forcings responsible for the initiation of the top-down mechanism [32]. For the fourth experiment (fixVIS/IR) we kept solar irradiance in the 250 to 4000 nm band at the 1851 level. This experiment helped to elucidate the role of a direct influence of solar irradiance on the troposphere and surface. For the fifth simulation (fixGHG), we use well-mixed greenhouses gases (CO_2, N_2O, and CH_4), ozone-depleting substances, and ozone precursor (NO_x and CO) emissions fixed at the 1851 level. The sixth simulation (fixWMGHG) was identical to fixGHG, except that NO_x and CO emissions were not fixed. Finally, the last simulation (noVOL) was performed prescribing the stratospheric aerosols at 1851 levels, which is typical for low volcanic activity. All experiments are listed in Table 1. The trend analysis was carried out for the ALL experiment applying a robust linear trend calculation for the 1910 to 1940 period with the nonparametric Sen–Mann–Kenndall trend significance test using a 90% confidence interval. We concentrated on this period to exclude the potential influence of a powerful tropical volcanic eruption in the 1902.

Table 1. The list of performed ensemble numerical experiments.

Experiment Name	Fixed Forcing	Color Code for Evolution Plots
ALL	None	Black
noEPP	Energetic particles	Violet
fixUV	Solar UV irradiance (λ < 250 nm), extra heating, and photolysis rates	Magenta
fixVIS/IR	Solar visible and near infrared irradiance	Light blue
fixGHG	CO_2, N_2O, CH_4, NO_x, and CO emissions	Green
fixWMGHG	CO_2, N_2O, and CH_4	Orange
noVOL	Stratospheric sulfate aerosol	Grey

3. Results

3.1. *Analysis of the Ozone Layer Evolution Drivers*

The evolution of the ozone layer is driven by a multitude of factors such as atmospheric circulation, transport, temperature, and concentration of ozone destroying reactive species [33], which in turn depend on the natural and anthropogenic forcing agents. The relative role of these drivers is difficult to elucidate from observational data, but the application of the model makes it possible.

3.1.1. Active Hydrogen Oxides

The active hydrogen oxides or HO_x ($OH + HO_2$) catalytically destroy ozone in the atmosphere. They are more effective in the lower and upper stratosphere [34]. Hydrogen oxides are produced from water vapor via photolysis by solar UV in the Lyman-α line and oxygen Schumann–Runge bands, or by reaction with exited atomic oxygen [33]. The solar activity modulates both factors because exited atomic oxygen is also produced by ozone photolysis. Water vapor depends on atmospheric transport and methane abundance, which are both modulated by natural and anthropogenic activities. The annual and zonal mean HO_x trend is illustrated in Figure 1.

The HO_x trend is most pronounced (more than 10%) above 50 km and in the tropical troposphere. To identify the drivers of such changes we show the evolution of the HO_x mixing ratio in Figure 2 for these two regions.

Figure 2a shows that, in the upper atmosphere above 50 km, the influence of solar UV irradiance dominates, because for the fixUV model experiment the HO_x mixing ratio does not change with time. For all other experiments, when the solar UV is not fixed, the HO_x mixing ratio follows the behavior of solar activity. In the troposphere (Figure 2b), anthropogenic emissions play the most important role. The HO_x increase is explained by enhanced water vapor in the warmer climate [19] and increased tropospheric ozone (Section 3.2). The direct sink of hydroxyl caused by enhanced methane and CO abundances is less important in the free troposphere, however, it almost completely eliminates the HO_x increase over the Northern Hemisphere, where the intensification of anthropogenic carbon monoxide emissions is most pronounced.

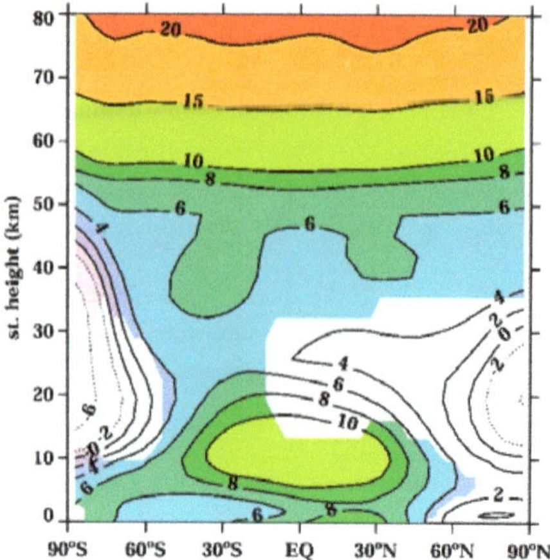

Figure 1. The annual and zonal mean HO_x linear trend (%/31 years) during the 1910 to 1940 period for the ALL experiment. The area where the trend is significant at the 90% or better level is marked by color shading.

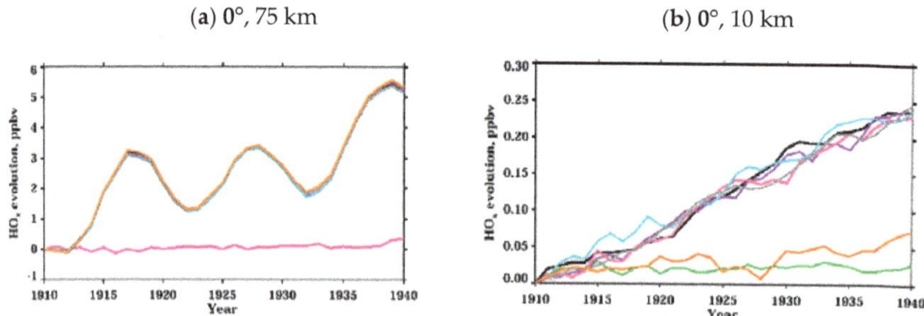

Figure 2. The time evolution of annual and zonal mean HO_x since 1910. (**a**) In the tropical upper mesosphere panel; and (**b**) in the equatorial middle troposphere panel. The lines are black for ALL, violet for noEPP, magenta for fixUV, light blue for fixVIS/IR, green for fixGHG, orange for fixWMGHG, and grey for noVOL experiments.

3.1.2. Nitrogen Oxides

Nitrogen oxides or NO_y ($N + NO + NO_2 + HNO_3 + HNO_4 + 2^*N_2O_5$) participate in catalytic ozone loss in the atmosphere. They are more effective in the middle stratosphere [34]. Nitrogen oxides are produced mostly from the N_2O reaction with exited atomic oxygen [33]. The precipitating energetic particles also produce NO_y over high latitudes during the winter season [35]. Solar activity modulates both factors because the concentration of exited atomic oxygen depends on ozone photolysis, and energetic particle precipitation depends on the solar wind. The main stratospheric loss of NO_y is the cannibalistic reaction $N + NO = N_2 + O$ [33] driven by NO photolysis. The tropospheric NO_y level strongly depends on anthropogenic activity [10]. The annual and zonal mean NO_y trend is illustrated in Figure 3.

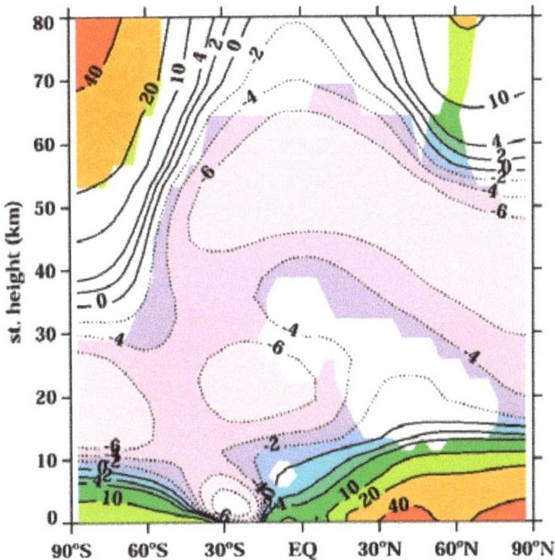

Figure 3. The annual and zonal mean NO_y linear trend (%/31 years) during the 1910 to 1940 period for the ALL experiment. The area where the trend is significant at the 90% or better level is marked by color shading.

The NO_y trend is most pronounced in the stratosphere, free northern troposphere, and polar mesosphere. The evolution of the NO_y mixing ratio is illustrated in Figure 4 for two above-mentioned regions to identify the responsible forcing.

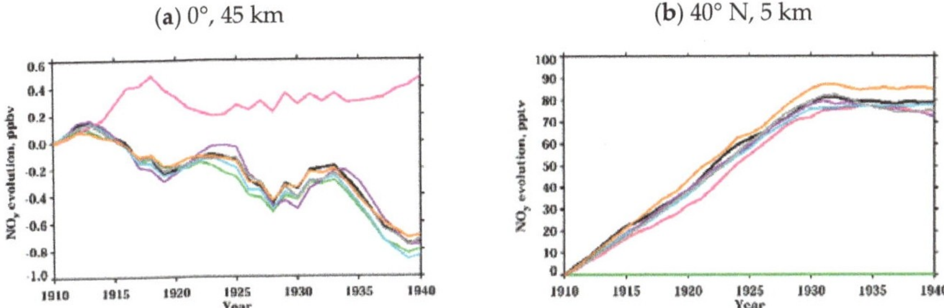

Figure 4. The evolution of annual and zonal mean NO_y since 1910. (**a**) In the tropical upper stratosphere panel; (**b**) in the lower troposphere over northern midlatitudes panel. The lines are black for ALL, violet for noEPP, magenta for fixUV, light blue for fixVIS/IR, green for fixGHG, orange for fixWMGHG, and grey for noVOL experiments.

Figure 4a demonstrates that the influence of solar UV irradiance dominates in the stratosphere, because without the UV forcing the NO_y mixing ratio only changes slightly with time. When the solar UV forcing is switched on, the behavior of the NO_y mixing ratio resembles the solar activity evolution due to the NO photolysis modulation by the solar activity. In the troposphere, the anthropogenic emissions of NO and NO_2 (NO_x) play the most important role leading to the substantial (by up to 90 pptv) NO_y increase. The results of the model run with fixed NOx emissions, shown by the green line in Figure 4b, demonstrate the absence of detectable NO_y evolution when the anthropogenic emissions of NO and NO_2 (NO_x) are fixed. Weak changes in NO_y after 1930 are related to a slower increase in anthropogenic NOx emissions [31]. The NO_y increase in the mesosphere is fully defined by the increased intensity of energetic particle precipitation caused by stronger solar and geomagnetic activity (not shown).

3.1.3. Temperature

Temperature is important for the processes regulating ozone balance, because kinetic reaction rates are temperature dependent [33]. Atmospheric temperature depends on a multitude of physical and chemical processes driven by both natural and anthropogenic factors. Figure 5 demonstrates temperature changes during the 1910 to 1940 period.

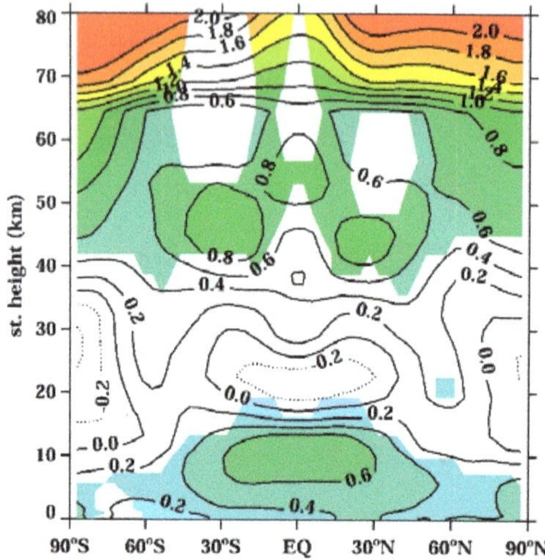

Figure 5. The annual and zonal mean temperature linear trend (K/31 years) during the 1910 to 1940 period for the ALL experiment. The area where the trend is significant at the 90% or better level is marked by color shading.

A warming trend is visible in almost the entire atmosphere but is small and not statistically significant in the middle stratosphere. The troposphere becomes warmer, in 1940, by up to 0.6 K as compared with in 1910. The obtained surface warming was described in detail by [19]. They obtained about 0.4 K global mean warming defined mostly by well-mixed greenhouse gases (about 50%) and solar irradiance in the visible and near infrared parts of the spectrum (35%). Some contribution (about 15%) comes from the tropospheric ozone increase caused by ozone precursor emissions. Stronger warming (1 to 2 K) appears in the upper stratosphere and mesosphere. Two warming spots in the upper stratosphere over the northern and southern tropics have the same origin as in the mesosphere, but of a smaller magnitude. Over the equator the temperature changes are smaller and have only marginal significance because the solar UV forcing is less efficient in this area and does not completely dominate over the greenhouse gas forcing.

Figure 6 illustrates the contribution of different factors to the temperature evolution during the period considered. Warming in the mesosphere, for the case with all drivers switched on (black line), is formed by competition between solar UV irradiance heating and cooling by well-mixed greenhouse gases with a small contribution from solar irradiance in the visible spectral region (light blue line in Figure 6a). When the solar UV irradiance is fixed (magenta line in Figure 6a), greenhouse gases cools mesosphere down by up to 2 K. In the absence of greenhouse gas changes (green and orange curves in Figure 6a), the heating from the absorption of enhanced solar UV irradiance leads to a warming of the mesosphere by up to 3 K. In addition, variable solar UV irradiance effects are visible in decadal scale variability with the magnitude of 0.75 K from the solar activity cycle.

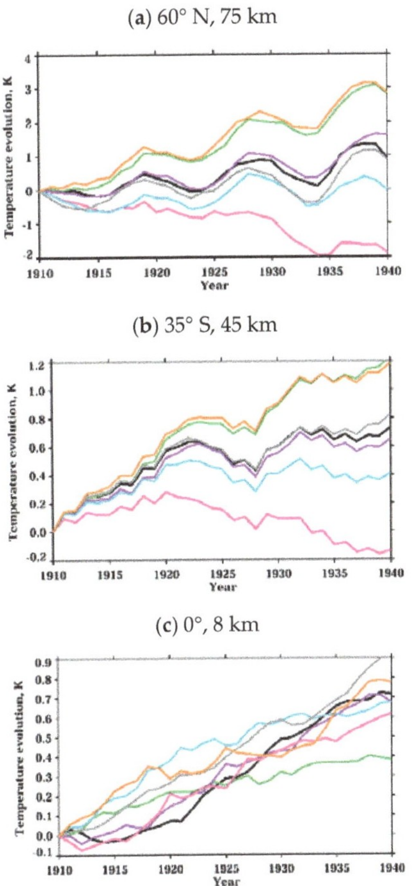

Figure 6. The evolution of annual and zonal mean temperature since the 1910. (**a**) In the northern midlatitude mesosphere panel; (**b**) in the upper stratosphere over southern midlatitudes panel; and (**c**) in the tropical middle troposphere panel. The lines are black for ALL, violet for noEPP, magenta for fixUV, light blue for fixVIS/IR, green for fixGHG, orange for fixWMGHG, and grey for noVOL experiments.

In the middle tropical troposphere, the analysis is rather complicated. Except the dominating contribution from tropospheric ozone precursors over well-mixed greenhouse gases (compare orange and green lines in Figure 6c), it is difficult to identify the most important factors.

3.2. Analysis of the Ozone Layer Evolution

The ozone changes depend on all drivers considered in Section 3.1, as well as on the transport processes related to continuous climate warming during the considered period [19]. Figure 7 demonstrates the ozone changes between 1910 and 1940. The obtained results allow three major areas with different ozone behavior to be identified as follows: the mesosphere, the middle stratosphere, and the troposphere. Ozone depletion is visible in the entire mesosphere, with the maxima in the polar regions of up to 10%. The opposite effect occurs in the middle stratosphere and troposphere, where the ozone concentration increases up to 5% and 20%, respectively. In the lower and upper stratosphere ozone trends are small and statistically insignificant.

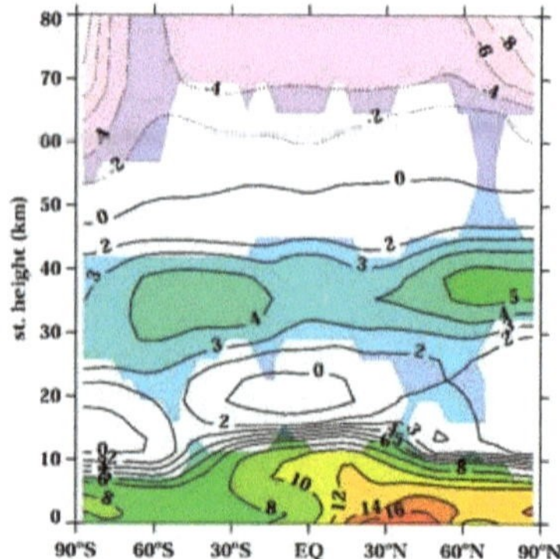

Figure 7. The annual and zonal mean ozone linear trend (%/31 years) during 1910 to 1940 for the ALL experiment. The area where the trend is significant at the 90% or better level is marked by color shading.

The time evolution of the annual zonal mean ozone mixing ratio and the contribution of different forcings for key altitude and latitude areas are shown in Figure 8. In the southern polar upper mesosphere (Figure 8a), the ozone evolution is reversed relative to solar activity for all experiments but has different magnitudes. The greatest contribution to the negative ozone trend in the mesosphere is related to the energetic particles, which produce more reactive hydrogen and nitrogen oxides during high solar and geomagnetic activity. The other drivers do not significantly affect the ozone evolution in this atmospheric region. The solar cycle is visible even in the absence of UV variability because the energetic particles forcing also has a decadal scale variability.

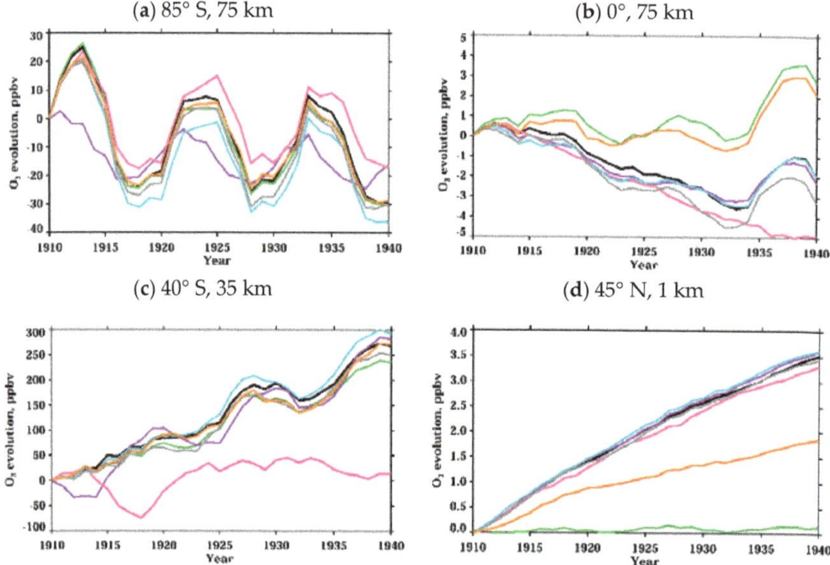

Figure 8. The evolution of annual zonal mean ozone mixing ratios since 1910. (**a**) In the southern polar upper mesosphere panel; (**b**) in the tropical mesosphere panel; (**c**) in the middle stratosphere over southern midlatitudes panel; and (**d**) in the the lower troposphere over northern midlatitudes panel. The lines are black for ALL, violet for noEPP, magenta for fixUV, light blue for fixVIS/IR, green for fixGHG, orange for fixWMGHG, and grey for noVOL experiments.

In the tropical mesosphere (Figure 8b) all drivers of ozone evolution can be separated into three groups. For the fixUV case, the ozone mixing ratio steadily decreases with time due to an increase in HO_x (Figure 2) caused by an increase in the production of water vapor from enhanced methane emission. The cooling of the mesosphere caused by the increase in well-mixed greenhouse gases (see the discussion of Figure 6a) suppresses the intensity of the ozone destruction cycles and leads to a small ozone increase, however, it cannot compensate for the influence of HO_x. The lack of a variable solar UV irradiance also explains the absence of cyclical ozone behavior which is visible for all other cases. For the fixGHG and fixWMGHG cases, the gradual ozone increase is driven mostly by the solar UV changes, which more than compensate for the HO_x increase due to the enhanced H_2O photolysis (Figure 2). Thus, the ozone evolution of the ALL experiment is formed from the competition between a greenhouse gas induced HO_x increase and a solar UV irradiance enhancement.

In the middle stratosphere over southern middle latitudes (Figure 8c), variations in solar UV irradiance play a dominant role that lead to a substantial increase in the ozone mixing ratio and almost constant values in the case when solar UV irradiance is fixed (case fixUV). In the lower troposphere over northern latitudes (Figure 8d), the ozone evolution is driven by tropospheric ozone precursors (fixGHG case) and, to a lesser extent, by the well-mixed greenhouse gases (fixWMGHG case). In the latter case, the ozone increase is related to enhanced methane emissions. The geographical distribution of the changes in annual mean total ozone from 1910 to 1940 driven by all considered forcing agents is illustrated in Figure 9. The simulated total ozone changes are positive and statistically significant all over the globe except in the western Pacific. In the tropical and high latitude belts, the changes are about 6 DU. More pronounced total column ozone trends are found over the middle latitudes in both hemispheres. There, the ozone change from 1910 to 1940.

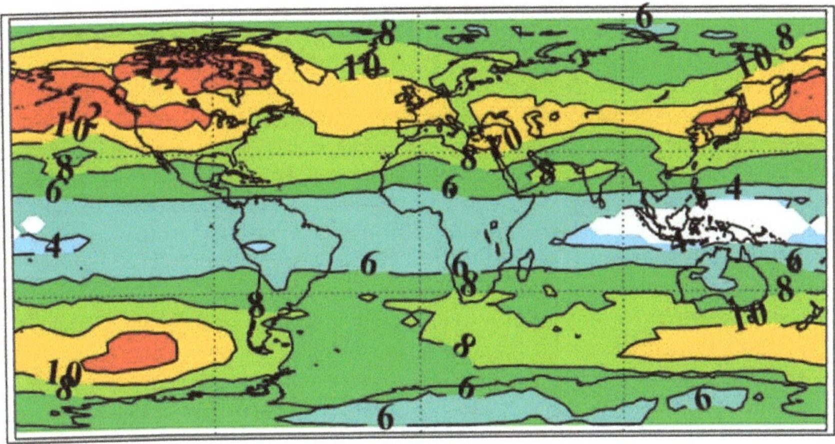

Figure 9. Geographical distribution of the annual mean total column ozone trend (DU/31 years) during.

The 1910 to 1940 time period from the run ALL, the area where the trend is significant at the 90% or better level, is marked by color shading and reaches 12 DU (about 4%) over North America, as well as over the northern and southern parts of the Pacific Ocean. Over Europe, the total ozone increase is slightly smaller but still exceeds 10 DU. These areas are typical locations of spring maxima in the total column ozone distribution caused by a spring-time acceleration of the meridional circulation transporting ozone down from its production area. Therefore, these changes in the total column ozone can be largely attributed to an increase in stratospheric production by enhanced solar UV irradiance. The contribution from tropospheric ozone is small because about 90% of the total column ozone is located in the stratosphere.

4. Discussion

Our results show the importance of an accurate prescription of both natural and anthropogenic forcings to reconstruct past climate and ozone layer trends. The estimate of the total column ozone trend during the first half of the 20th century, obtained by [15] using mostly anthropogenic forcing, does not exceed 1.0 DU. A consideration of the natural forcing by [16] led to the much higher value of almost 3.0 DU. In the present work, using new estimates for the solar forcing from [28], we obtained a three times stronger effect reaching almost 8 DU for the global annual mean value, and 12 DU over the northern and southern middle latitudes. The enhanced magnitude of the total column ozone changes is explained solely by the applied strong solar forcing. Thus, a poor understanding of the solar forcing automatically leads to large uncertainty in the simulated total ozone evolution. However, a high sensitivity of the total column ozone to solar UV forcing can be used to resolve long standing issues about the absolute value of the magnitude of past solar irradiance variability discussed recently by [17]. This problem can potentially be solved by comparing the simulated total ozone behavior with direct measurements or proxy-based reconstructions. Unfortunately, direct comparisons of simulated and observed total column ozone trends is not possible from 1910 to 1940 because the observations are available only from 1926 [36]. The proxy-based reconstructions of total ozone are not available at the moment, but there is some progress in this direction. One possible approach is to retrieve surface UV-B radiation level at the surface from the analysis of UV-B absorbing objects in the plants or spores [37–40]. However, it is not clear how to separate the influence of stratospheric ozone variations from the spectral solar irradiance variability, which is not well constraint on long-term time scales [17]. The simulated changes in the total column ozone are mainly caused by the ozone evolution in the middle stratosphere driven by steadily growing extraterrestrial solar UV irradiance. The simulated shape of the stratospheric ozone increase, shown in Figure 7, resembles the ozone response to the solar UV irradiance enhancement during the recent period [40] only in the tropical middle stratosphere (between 30 and 40 km). The elevated secondary ozone enhancement in the upper stratosphere over midlatitudes obtained by [40] from observations and model simulations are not visible in our results. The difference can be explained by different circulation regimes in these two periods. A possible influence from the circulation is illustrated in [40] from a comparison of free running and specified dynamics model runs. In the case of a free running model, the upper stratospheric spots of enhanced ozone are much less pronounced in comparison with specified dynamics runs. This difference should be related to different circulation fields, because the treatment of chemical and transport processes is identical in both model versions.

An accurate knowledge of the tropospheric ozone evolution is also important, because it can play an important role in the explanation of the early 20th century warming (ETCW). It was shown in [19] that tropospheric ozone precursors (CO and NO_X) are the third most important factor influencing climate during this period, after greenhouse gases, solar visible, and infrared radiation. Therefore, an underestimation of the CO and NO_X emission intensification can explain an underestimation of the ETCW magnitude in many climate models [18]. The simulated annual mean tropospheric ozone mixing ratio in 1910 varies between 15 and 30 ppb over the northern mid-latitudes depending on the location and season, which overestimates 10 to 15 ppb obtained from direct surface ozone measurements at different locations in central Europe [6,10]. It should be noted, however, that these historical measurements probably underestimate ozone mixing ratios due to interference from water vapor and other species [41]. The subsequent ozone increase (Figures 7 and 8d) of about 4 ppb (~15%) during 1910 to 1940, in our experiment, is close to the 11% increase simulated by [10], and 16% obtained from direct ozone measurements at mountain sites [5]. Our simulations of the rate of the tropospheric ozone increase agree with the isotope analysis of air trapped in the ice and snow, as well as with results from the GISS-E2.1 model [41].

The presented analysis can be extended to cover trends of halogenated species and atmospheric dynamics. We do not expect substantial contributions from the halogenated species because of their very small (more than six times as compare with present day) abundance and absence of trends during the considered period. Dynamical changes caused by climate warming could consist of an altered tropopause height, state of the polar vortices, or Brewer–Dobson circulation (BDC) intensity. However, the analysis of their trends is more difficult because the response of dynamical properties has a low signal-to-noise ratio. For example, the lower stratospheric ozone depletion in the tropical area caused by enhanced BDC in a warmer climate, and clearly visible in the simulation of the future climate [42], is not significant in our case (Figure 7). The same can be said about the dipole-like polar temperature changes which characterize polar vortex strengthening (Figure 5). Probably, these changes should be examined in the future on seasonal or even monthly time scales.

5. Conclusions

In this study of ozone layer evolution during the early 20th century, we exploited the chemistry-climate model SOCOL-MPIOM driven by all known anthropogenic and natural forcing agents, as well as their combinations. Using results from seven ten-member ensemble runs, we demonstrate the time evolution of the main factors responsible for t ozone production and loss from the ground to the mesopause. We demonstrate that in the mesosphere the ozone mixing ratio trend during the 1910 to 1940 period is negative and driven by energetic particles, incoming solar UV radiation, and greenhouse gases. In the middle stratosphere, the ozone increased from 1910 to 1940 by up to 5%, mostly due to the enhancement of solar UV radiation.

Our calculations emphasize the dominant role of anthropogenic factors in the troposphere, where an increase in CO and NO_x emissions leads to an increase in ozone mixing ratios by up to 15%. The general agreement of the increase in tropospheric ozone with previously published estimates allows us to conclude that a climate influence from this forcing is rather well constrained and cannot explain the underprediction of the ETCW magnitude by many climate models [18,19].

We obtained a significant global scale increase in the total column ozone exceeding 12 Dobson Units over northern and southern middle latitudes. We conclude that total column ozone changes during this period were driven mostly by an enhancement in solar UV radiation. Our simulation results can be used to constrain the solar forcing magnitude if past ozone or solar UV-B radiation becomes available.

Author Contributions: Data curation, T.S.; Software, P.A.; Supervision, E.R.; Writing - original draft, T.E. All authors have read and agreed to the published version of the manuscript.

Funding: This research was funded by the Swiss National Science Foundation under grants CRSII2-147659 (FUPSOL-II) and 200020_182239 (POLE).

Acknowledgments: Numerical simulations were performed on the ETH Zürich cluster EULER. Our special thanks to W. Ball for the English grammar improvement.

Conflicts of Interest: The authors declare no conflict of interest.

References

1. Bais, A.F.; Lucas, R.M.; Bornman, J.F.; Williamson, C.E.; Sulzberger, B.; Austin, A.T.; Wilson, S.R.; Andrady, A.L.; Bernhard, G.; McKenzie, R.L.; et al. Environmental effects of ozone depletion, UV radiation and interactions with climate change: UNEP Environmental Effects Assessment Panel, update 2017. *Photochem. Photobiol. Sci.* **2018**, *17*, 127–179. [CrossRef] [PubMed]
2. Häder, D.-P.; Barnes, P.W. Comparing the impacts of climate change on the responses and linkages between terrestrial and aquatic ecosystems. *Sci. Total Environ.* **2019**, *682*, 239–246. [CrossRef] [PubMed]
3. WMO (World Meteorological Organization). *Scientific Assessment of Ozone Depletion: 2018*; Global Ozone Research and Monitoring Project—Report No. 58; WMO: Geneva, Switzerland, 2018; 588p.
4. Volz, A.; Kley, D. Evaluation of the Montsouris series of ozone measurements made in the nineteenth century. *Nature* **1988**, *332*, 240–242. [CrossRef]
5. Marenco, A.; Gouget, H.; Nedelec, P.; Pages, J.-P. Evidence of a long-term increase in tropospheric ozone from Pic du Midi data series: Consequences: Positive radiative forcing. *J. Geophys. Res.* **1994**, *99*, 16617–16632. [CrossRef]

6. Hauglustaine, D.A.; Brasseur, G.P. Evolution of tropospheric ozone under anthropogenic activities and associated radiative forcing of climate. *J. Geophys. Res.* **2001**, *106*, 32337–32360. [CrossRef]
7. Chalita, S.; Hauglustaine, D.A.; Letreut, H.; Müller, J.-F. Radiative forcing due to increased tropospheric ozone concentrations. *Atmos. Environ.* **1996**, *30*, 1641–1646. [CrossRef]
8. Mickley, L.J.; Jacob, D.J.; Rind, D. Uncertainty in preindustrial abundance of tropospheric ozone: Implications for radiative forcing calculations. *J. Geophys. Res.* **2001**, *106*, 3389–3399. [CrossRef]
9. Shindell, D.T.; Faluvegi, G.; Bell, N. Preindustrial-to-present-day radiative forcing by tropospheric ozone from improved simulations with the GISS chemistry-climate GCM. *Atmos. Chem. Phys.* **2003**, *3*, 1675–1702. [CrossRef]
10. Lamarque, J.-F.; Hess, P.; Emmons, L.; Buja, L.; Washington, W.; Granier, C. Tropospheric ozone evolution between 1890 and 1990. *J. Geophys. Res.* **2004**, *110*, D08304. [CrossRef]
11. Parrella, J.P.; Jacob, D.J.; Liang, Q.; Zhang, Y.; Mickley, L.J.; Miller, B.F.; Evans, M.J.; Yang, X.; Pyle, J.A.; Theys, N.; et al. Tropospheric bromine chemistry: implications for present and pre-industrial ozone and mercury. *Atmos. Chem. Phys.* **2012**, *12*, 6723–6740. [CrossRef]
12. Young, P.J.; Archibald, A.T.; Bowman, K.W.; Lamarque, J.-F.; Naik, V.; Stevenson, D.S.; Tilmes, S.; Voulgarakis, A.; Wild, O.; Bergmann, D. Pre-industrial to end 21st century projections of tropospheric ozone from the Atmospheric Chemistry and Climate Model Intercomparison Project (ACCMIP). *Atmos. Chem. Phys.* **2013**, *13*, 2063–2090. [CrossRef]
13. Reader, M.C.; Plummer, D.A.; Scinocca, J.F.; Shepherd, T.G. Contributions to twentieth century total column ozone change from halocarbons, tropospheric ozone precursors, and climate change. *Geophys. Res. Lett.* **2013**, *40*, 6276–6281. [CrossRef]
14. Hollaway, M.J.; Arnold, S.R.; Collins, W.J.; Folberth, G.; Rap, A. Sensitivity of midnineteenth century tropospheric ozone to atmospheric chemistry-vegetation interactions. *Geophys. Res. Atmos.* **2011**, *122*, 2452–2473. [CrossRef]
15. Fleming, E.L.; Jackman, C.H.; Stolarski, R.S.; Douglass, A.R. A model study of the impact of source gas changes on the stratosphere for 1850–2100. *Atmos. Chem. Phys.* **2011**, *11*, 8515–8541. [CrossRef]
16. Lean, J. Evolution of Total Atmospheric Ozone from 1900 to 2100 Estimated with Statistical Models. *J. Atmos. Sci.* **2014**, *71*, 1956–1984. [CrossRef]
17. Egorova, T.; Schmutz, W.; Rozanov, E.; Shapiro, A.I.; Usoskin, I.; Beer, J.; Tagirov, R.V.; Peter, T. Revised historical solar irradiance forcing. *Astron. Astrophys.* **2018**, *615*, A85. [CrossRef]
18. Hegerl, G.C.; Brönnimann, S.; Schurer, A.; Cowan, T. The early 20th century warming: anomalies, causes, and consequences. *WIREs Clim. Chang.* **2018**, *9*, e522. [CrossRef]
19. Egorova, T.; Rozanov, E.; Arsenovic, P.; Peter, T.; Schmutz, W. Contributions of Natural and Anthropogenic Forcing Agents to the Early 20th Century Warming. *Front. Earth Sci.* **2018**, *6*, 206. [CrossRef]
20. Muthers, S.; Anet, J.G.; Stenke, A.; Raible, C.C.; Rozanov, E.; Brönnimann, S.; Peter, T.; Arfeuille, F.X.; Shapiro, A.I.; Beer, J.; et al. The coupled atmosphere-chemistry-ocean model SOCOL MPIOM. *Geosci. Model. Dev.* **2014**, *7*, 2157–2179. [CrossRef]
21. Arsenovic, P.; Rozanov, E.; Anet, J.; Stenke, A.; Peter, T. Implications of potential future grand solar minimum for ozone layer and climate. *Atmos. Chem. Phys.* **2018**, *18*, 3469–3483. [CrossRef]
22. Stenke, A.; Schraner, M.; Rozanov, E.; Egorova, T.; Luo, B.; Peter, T. The SOCOL version 3.0 chemistry–climate model: description, evaluation, and implications from an advanced transport algorithm. *Geosci. Model. Dev.* **2013**, *6*, 1407–1427. [CrossRef]
23. Roeckner, E.; Bäuml, G.; Bonaventura, L.; Brokopf, R.; Esch, M.; Giorgetta, M.; Hagemann, S.; Kirchner, I.; Kornblueh, L.; Manzini, E.; et al. *The Atmospheric General Circulation Model ECHAM5. Part I: Model Description*; Report No. 349; Max Planck Institute for Meteorology: Hamburg, Germany, 2003.
24. Rozanov, E.V.; Zubov, V.; Schlesinger, M.E.; Yang, F.; Andronova, N.G. The UIUC three-dimensional stratospheric chemical transport model: description and evaluation of the simulated source gases and ozone. *J. Geophys. Res.* **1999**, *104*, 11755–11781. [CrossRef]
25. Egorova, T.; Rozanov, E.; Zubov, V.; Karol, I. Model for investigating ozone trends (MEZON). *Izv. Atmos. Ocean. Phys.* **2003**, *39*, 277–292.
26. Marsland, S.J.; Haak, H.; Jungclaus, J.H.; Latif, M. The Max-Planck-Institute global ocean/sea ice model with orthogonal curvilinear coordinates. *Ocean Model.* **2003**, *5*, 91–127. [CrossRef]

27. Jungclaus, J.H.; Keenlyside, N.; Botzet, M.; Haak, H.; Luo, J.-J.; Latif, M.; Marotzke, J.; Mikolajewicz, U.; Roeckner, E. Ocean Circulation and Tropical Variability in the Coupled Model ECHAM5/MPI-OM. *J. Clim.* **2006**, *19*, 3952–3972. [CrossRef]
28. Shapiro, A.I.; Schmutz, W.; Rozanov, E.; Schoell, M.; Haberreiter, M.; Shapiro, A.V.; Nyeki, S. A new approach to the long-term reconstruction of the solar irradiance leads to large historical solar forcing. *Astron. Astrophys.* **2011**, *529*, A67. [CrossRef]
29. Sukhodolov, T.; Rozanov, E.; Shapiro, A.I.; Anet, J.; Cagnazzo, C.; Peter, T.; Schmutz, W. Evaluation of the ECHAM family radiation codes performance in representation of the solar signal. *Geosci. Model. Dev.* **2014**, *7*, 2859–2866. [CrossRef]
30. Matthes, K.; Funke, B.; Andersson, M.E.; Barnard, L.; Beer, J.; Charbonneau, P.; Clilverd, M.A.; de Wit, T.D.; Haberreiter, M.; Hendry, A.; et al. Solar forcing for CMIP6 (v3. 2). *Geosci. Model. Dev.* **2017**, *10*, 2247–2302. [CrossRef]
31. Meinshausen, M.; Smith, S.J.; Calvin, K.; Daniel, J.S.; Kainuma, M.L.T.; Lamarque, J.-F.; Matsumoto, K.; Montzka, S.A.; Raper, S.C.B.; Riahi, K.; et al. The RCP greenhouse gas concentrations and their extensions from 1765 to 2300. *Clim. Chang.* **2011**, *109*, 213–241. [CrossRef]
32. Gray, L.J.; Beer, J.; Geller, M.; Haigh, J.D.; Lockwood, M.; Matthes, K.; Cubasch, U.; Fleitmann, D.; Harrison, R.G.; Hood, L.; et al. Solar influence on climate. *Rev. Geophys.* **2010**, *48*, RG4001. [CrossRef]
33. Brasseur, G.P.; Solomon, S. *Aeronomy of the Middle Atmosphere: Chemistry and Physics of the Stratosphere and Mesosphere*; Springer: Dordrecht, The Netherlands, 2005; Volume 32, p. 646. [CrossRef]
34. Larin, I. On the chain length and rate of ozone depletion in the main stratospheric cycles. *Atmos. Clim. Sci.* **2013**, *3*, 141–149. [CrossRef]
35. Mironova, I.A.; Aplin, K.L.; Arnold, F.; Bazilevskaya, G.A.; Harrison, R.G.; Krivolutsky, A.A.; Nicoll, K.A.; Rozanov, E.V.; Turunen, E.; Usoskin, I.G. Energetic Particle Influence on the Earth's Atmosphere. *Space Sci. Rev.* **2015**, *194*, 1–96. [CrossRef]
36. Scarnato, B.; Staehelin, J.; Stübi, R.; Schill, H. Long-term total ozone observations at Arosa (Switzerland) with Dobson and Brewer instruments (1988–2007). *J. Geophys. Res.* **2010**, *115*, D13306. [CrossRef]
37. Rozema, J.; van Geel, B.; Bjorn, L.O.; Lean, J.; Madronich, S. PALEOCLIMATE: Toward Solving the UV Puzzle. *Science* **2002**, *296*, 1621–1622. [CrossRef]
38. Magri, D. Past UV-B flux from fossil pollen: Prospects for climate, environment and evolution. *New Phytol.* **2011**, *192*, 310–312. [CrossRef] [PubMed]
39. Jardine, P.E.; Fraser, W.T.; Gosling, W.D.; Roberts, C.N.; Eastwood, W.J.; Lomax, B.H. Proxy reconstruction of ultraviolet-B irradiance at the Earth's surface, and its relationship with solar activity and ozone thickness. *Holocene* **2019**. [CrossRef]
40. Ball, W.T.; Rozanov, E.; Alsing, J.; Marsh, D.R.; Tummon, F.; Mortlock, D.J.; Kinnison, D.; Haigh, J.D. The upper stratospheric solar cycle ozone response. *Geophys. Res. Lett.* **2019**, *46*, 1831–1841. [CrossRef]
41. Yeung, L.Y.; Murray, L.T.; Martinerie, P.; Witrant, E.; Hu, H.; Banerjee, A.; Orsi, A.; Chappellaz, J. Isotopic constraint on the twentieth-century increase in tropospheric ozone. *Nature* **2019**, *570*, 224–227. [CrossRef]
42. Zubov, V.; Rozanov, E.; Egorova, T.; Karol, I.; Schmutz, W. Role of external factors in the evolution of the ozone layer and stratospheric circulation in 21st century. *Atmos. Chem. Phys.* **2013**, *13*, 4697–4706. [CrossRef]

Article

Effects of Ozone and Clouds on Temporal Variability of Surface UV Radiation and UV Resources over Northern Eurasia Derived from Measurements and Modeling

Natalia E. Chubarova [1,*], Anna S. Pastukhova [1], Ekaterina Y. Zhdanova [1], Elena V. Volpert [1], Sergey P. Smyshlyaev [2] and Vener Y. Galin [3]

1 Faculty of Geography, Lomonosov Moscow State University, GSP-1, Leninskie Gory, 1119991 Moscow, Russia; p-annet@mail.ru (A.S.P.); ekaterinazhdanova214@gmail.com (E.Y.Z.); elena.volpert@yandex.ru (E.V.V.)
2 Russian State Hydrometeorological University, Malookhtinsky Ave., 98, 195196 Saint-Petersburg, Russia; smyshl@rshu.ru
3 Institute of Numerical Mathematics RAS, Gubkina str., 8, 119991 Moscow, Russia; venergalin@yandex.ru
* Correspondence: natalia.chubarova@gmail.com

Received: 25 November 2019; Accepted: 23 December 2019; Published: 2 January 2020

Abstract: Temporal variability in erythemal radiation over Northern Eurasia (40°–80° N, 10° W–180° E) due to total ozone column (X) and cloudiness was assessed by using retrievals from ERA-Interim reanalysis, TOMS/OMI satellite measurements, and INM-RSHU chemistry–climate model (CCM) for the 1979–2015 period. For clear-sky conditions during spring and summer, consistent trends in erythemal daily doses (*Eery*) up to +3%/decade, attributed to decreases in X, were calculated from the three datasets. Model experiments suggest that anthropogenic emissions of ozone-depleting substances were the largest contributor to *Eery* trends, while volcanic aerosol and changes in sea surface temperature also played an important role. For all-sky conditions, *Eery* trends, calculated from the ERA-Interim and TOMS/OMI data over the territory of Eastern Europe, Siberia and Northeastern Asia, were significantly larger (up to +5–8%/decade) due to a combination of decrease in ozone and cloudiness. In contrast, all-sky maximum trends in *Eery*, calculated from the CCM results, were only +3–4%/decade. While *Eery* trends for Northern Eurasia were generally positive, negative trends were observed in July over central Arctic regions due to an increase in cloudiness. Finally, changes in the ultraviolet (UV) resources (characteristics of UV radiation for beneficial (vitamin D production) or adverse (sunburn) effects on human health) were assessed. When defining a "UV optimum" condition with the best balance in *Eery* for human health, the observed increases in *Eery* led to a noticeable reduction of the area with UV optimum for skin types 1 and 2, especially in April. In contrast, in central Arctic regions, decreases in *Eery* in July resulted in a change from "UV excess" to "UV optimum" conditions for skin types 2 and 3.

Keywords: total ozone content; cloudiness; erythemal radiation; trend; chemical–climate model; ERA-Interim reanalysis; Northern Eurasia; UV resources

1. Introduction

Ultraviolet (UV) radiation is well-known for its significant influence on human health and the environment. High UV doses have negative effects on skin (erythema (sunburn), skin cancer), and cause eye diseases and immune suppression [1,2]. However, moderate UV doses have positive effects causing vitamin D production [3,4]. Apart from the solar elevation, ozone and cloudiness are the main factors affecting UV level [2,5] and providing significant year-to-year variability of UV radiation. Erythemal radiation, which is used for characterizing UV health effects, is the UV radiation, weighted by the erythemal action spectrum with maximum efficiency at 280–300 nm and integrated over 280–400 nm [1].

As stated in numerous publications and assessments [5–8], the decrease in stratospheric ozone resulting from anthropogenic emissions of ozone depleting substances (ODSs) into the atmosphere has led to the increase in erythemal radiation at the Earth's surface in the 1980s and 1990s. Since the end of 1990s there is a small increase in total ozone content from 0.3% to 1.2% per decade due to the reduction of the ODSs in the stratosphere but this increase is not statistically significant [8]. However, there is no clear relationship between total ozone column (X) and erythemal radiation for many regions due to the influence of other factors like cloudiness, aerosols, minor gas species, and surface reflectivity [9,10]. The most important atmospheric factor is cloudiness, which significantly affects erythemal radiation and can enhance or mask its trends due to ozone changes.

There are a lot of studies devoted to the temporal and spatial variability of erythemal radiation over the globe or large spatial areas using the retrievals from satellite instruments, and model simulations [9–11], and references therein]. According to these data, different kinds of UV climatologies at various spatial scale have been obtained [12–15]. In addition to the UV climatologies, many publications are devoted to the analysis of temporal changes in *Eery* due to ozone and cloudiness. In [16] the analysis of satellite data over the globe has revealed an increase in erythemal radiation of about 8% at 50° S and of about 5% at 50° N over the 1978–2008 period due to decrease in X, while the combined effect of ozone and cloudiness led to the increase of about 2% at 50° S and 6% at 50° N. The changes in erythemal radiation, evaluated over the 1979–2010 period using TOMS/OMI retrievals, have also shown positive trends up to 5% at 45°–50° S in spring–summer seasons [17]. A statistically significant decrease in Lambertian Equivalent Reflectivity due to clouds (3.6 ± 0.02%) over the 1979–2011 period was derived from different satellite measurements [18,19]. However, satellite data have a large uncertainty due to problems with accounting for absorbing aerosol and due to the restrictions in UV retrievals over bright snow/ice surfaces [20,21] compared with direct ground-based measurements.

Many publications are devoted to the analysis of temporal variations in UV radiation obtained over local sites ([2] and references therein, [22–33]). According to these publications, an increase of UV radiation has been revealed over many European sites since the middle of 20 century, mainly due to ozone and cloud effects [28,29,31,33]. Over several locations (e.g., Thessaloniki) the increase in *Eery* to some extent can be also attributed to the decrease in aerosol optical thickness [34]. However, in the west Arctic area at Hornsund (77°00′ N, 15°33′ E) no pronounced trend in *Eery* has been found since 1983 due to complicated changes in cloud cover, which is one of the basic drivers of the long-term UV changes over this region [30].

Over Europe the UV climatology and long-term variability in erythemal radiation was analyzed within the framework of the COST Action 726 project [32] using accurate model simulations, the results of ERA-40 reanalysis over the five-decade period since 1950, and ground-based measurements. These data also demonstrate pronounced variations in *Eery* due to ozone and cloudiness with some specific features over different areas.

For evaluating variations in *Eery* in the past and future, global Earth-system models from the CMIP-5 (Coupled Model Intercomparison Project Phase 5) experiment, the ensemble mean of the Community Earth System Model version 1 (CESM1) and the Whole Atmosphere Community Climate Model (WACCM) with interactive chemistry ensemble of model ozone predictions have been used for estimating the UV variability over the 1955–2100 period [5]. Using this approach, the effects of various factors on future UV variations were estimated. Chemistry–climate models (CCMs) have also been used for estimating changes in UV radiation in the past and future [9,11,35–38]. In [35] the projections of erythemal irradiance from 1960 to 2100 have been made using radiative transfer calculations and projections of ozone, temperature and cloud change from 14 CCM, as part of the CCMVal-2 activity of SPARC (Stratosphere-troposphere Processes And their Role in Climate) project. According to this study annually mean erythemal radiation in the 2090s relative to 1980 will be on average approximately 12% lower at high latitudes in both hemispheres, and 3% lower at midlatitudes. At northern high latitudes, the increase in cloudiness towards the end of the 21st century will reduce the erythemal irradiance by 5% with respect to the 1960s.

In Ref. [37] the analysis of *Eery* variations has been made by using the clear-sky data from the first phase of the Chemistry–Climate Model Initiative (CCMI) as input to the TUV (Tropospheric Ultraviolet and Visible) radiation model. This analysis projects an increase in average *Eery* of about 2–4% in 2100 in the tropical belt (30° N–30° S) and a 1.8% to 3.4% increase in the midlatitudes in the Southern Hemisphere for RCP (Representative Concentration Pathway) 2.6, 4.5 and 6.0, compared to 1960s, which partly contradict to the results obtained in [35]. The projected increase in *Eery* reported in [37] results from the assumption that the atmospheric aerosol loading will decrease greatly over the course of the 21st century, which is debatable. The analysis of erythemal radiation according to the CCMI simulations [2] projects *Eery* to decrease by 5–15% in the northern hemisphere during summer and autumn mainly due to ozone recovery in 2085–2095 compared with 2010–2020 according to the RCP 6.0 scenario. However, some other factors like cloudiness, aerosol and surface reflectivity can also play a significant role over several regions. In [39] negative *Eery* changes over Northern midlatitudes of about 6–8% due to ozone and clouds were found by the end of 21 century, which are in agreement with [35].

For a wide range of biological and medical studies, it is important to interpret long-term changes in UV radiation from the point of health effects. The approach proposed in [14] distinguishes health effects between harmful and beneficial using the term "UV resources". The UV resources characterize UV radiation from the point of beneficial (vitamin D production) or adverse (sunburn) effects on human health. Their classification is given in the categories of UV deficiency, UV optimum and UV excess and takes into account the skin type and the fraction of the skin surface that is exposed to UV radiation. A similar approach has been previously developed in [40]. Changes in UV resources over the 1979–2015 period were first simulated for Moscow conditions in [28], where a transition from the UV optimum to UV moderate excess conditions for the population with the most vulnerable skin type I has been obtained in spring.

The main aim of this study is to identify temporal variability in Eery and the change in UV resources due to ozone and cloudiness between 1979 and 2015 based on the TOMS/OMI satellite datasets, and UV retrievals from the reanalysis and a chemistry–climate model over Northern Eurasia (40° N–80° N, 10° W–180° E). We chose this period since it was characterized by the most pronounced changes in *Eery*, and the quality of data is the most reliable. In the analysis we used the updated TOMS and OMI satellite UV retrievals using the new Macv2 absorbing aerosol correction [41], as well as the ERA-Interim Reanalysis data [42]. For revealing the causes of changes in ozone we applied model runs of the Russian Chemistry–Climate Model (CCM), which was jointly developed by the Institute of Numerical Mathematics RAS (INM) and Russian State Hydrometeorological University (RSHU) [43,44].

2. Methods and the Data Description

Long-term variability in *Eery* over Northern Eurasia was estimated using different datasets, which include the UV data from model, reanalysis, and satellite measurements. For evaluating the UV variability and increasing the efficiency of the simulations, we estimated the relative anomalies V_i in *Eery* as a sum of anomalies due to total ozone amount ($v1_{i,j}$ (X)) and due to cloud variations ($v2_{i,j}$ (*Cl*)). The approach is appropriate, since ozone and cloudiness are located at different levels of the atmosphere. This simple additive method has been used in the UV reconstruction model described in [33], where a good agreement has been shown between results obtained with this method and the observed interannual *Eery* variations. It was also applied for reconstructing the long-term UV variability over Moscow in [10]. The total *Eery* anomalies V_i for year i can be estimated with the following equation:

$$V_i = \sum_j (W_j(h_e)\ (v1_{i,j}(X) + v2_{i,j}(Cl)) / \sum_j W_j(h_e), \quad (1)$$

where index i and j correspond to a year and a month number respectively. W_j (h_e) is a weighting function, which is calculated using the effective solar elevation h_e.

For an accurate estimation of the effective solar elevation we used simulations of h with 1-hour resolution over a month. An effective solar elevation for each grid was estimated only for hours with $h > 0$. The annual solar elevation changes were accounted in W (h_e), using a power law dependence ($Eery \sim h_e{}^{\alpha}$), where $\alpha = 2$ according to [45]. The same W (h_e) correction has been applied to all three datasets (INM-RSHU CCM, ERA-Interim, and TOMS/OMI). In our simulations, we calculated anomalies ($v1_{i,j}$ (X) and $v2_{i,j}$ (Cl)) relative to 2000.

To evaluate the interannual changes in *Eery* due to ozone, we used a well-known Radiation Amplification Factor (*RAF*) technique [46]:

$$log(Eery_{i,j}) = -RAF \times log(X_{i,j}) + C, \tag{2}$$

where C is a constant.

In addition, we took into account that the *RAF* for erythemal irradiance depends on solar elevation. We used the following equation according to [47] to parameterize this relationship:

$$RAF(h_e) = -1.10 \times 10^{-4} h_e{}^2 + 1.57 \times 10^{-2} h_e + 0.665 \tag{3}$$

As a result, variations $v1_{i,j}$ can be written in the following way:

$$v1_{i,j}(X) = \left(\frac{X_{i,j} - X_{2000,j}}{X_{2000,j}}\right)^{-RAF(h_e(j))}, \tag{4}$$

where the index 2000 indicates the year 2000, and, hence, $X_{2000,j}$ is a monthly total ozone column in 2000.

In order to characterize both cloud geometry and cloud optical properties effects on *Eery* we applied Cloud Modification Factor for UV spectral region (CMF_{UV}), which is often used for this purpose [5,38] and can be estimated from the standard outputs (downward shortwave radiation in clear-sky and in cloudy conditions) of different CCMs and reanalysis datasets. Using these data, one can easily evaluate shortwave radiation cloud modification factor (*CMF*):

$$CMF = Q/Q_0, \tag{5}$$

where Q is the surface shortwave radiation in cloudy sky conditions, and Q_0 is the surface shortwave radiation in the cloudless atmosphere.

The spectral correction of the cloud modification factor was obtained using accurate model simulations in UV and total shortwave spectrum according to the 8-stream discrete ordinate radiative transfer method in the TUV model [48] and Monte-Carlo simulations [49] for different atmospheric conditions and solar elevations. As a result, we showed that for optically thin cloudiness with $CMF > 0.95$ we can neglect its spectral dependence, while a correction should be used in other conditions. According to [28], this correction is the following:

$$CMF_{UV} = (0.1417 \times sin(h_e)^2 - 0.175 \times sin(h_e) + 1.054) CMF^{(-0.04 \times sin(h_e)^2 + 0.554 \times sin(h_e) + 0.609)} \tag{6}$$

The proposed method has been successfully tested against the long-term measurements of erythema radiation at the Moscow State University Meteorological Observatory over the 1999–2015 period [28].

To account for total ozone and cloud variations we applied the same method to the different datasets (INM-RSHU CCM, ERA-Interim reanalysis, satellite TOMS/OMI). We should note, that in this study we do not take into account changes in aerosol optical properties and surface reflectivity, which could have some direct and indirect radiative effect on variations in *Eery*, but at much smaller level comparing with ozone and cloudiness over the 1979–2015 period [28].

According to Equations (1) to (6) we evaluated anomalies in *Eery* relative to the year 2000 taking into account either only total ozone or only the cloud modification factor, and considering both of these parameters. After obtaining the *Eery* anomalies, the comparisons between the mean anomaly in *Eery* over the 2000–2015 period and over the 1979–1999 period were performed. For characterizing the statistical significance of the differences in *Eery* between the two periods, we applied a t-test for the means of two independent samples (Welch's *t*-test) at $p = 95\%$.

In addition, we performed a trend analysis for *Eery* anomalies using a linear regression model with the least squares approach and estimated its uncertainty using t-test at $p = 95\%$.

The details of the UV resources estimations are given in Section 2.4.

2.1. The Description of INM-RSHU Chemical Climate Model

The INM-RSHU global three-dimensional chemistry–climate model consists of two modules: a dynamical module, which has been developed at the Institute of Numerical Mathematics (INM) of Russian Academy of Science (RAS) [44] and a photochemical module developed at the Russian State Hydrometeorological University (RSHU). Both modules were successfully applied in different international projects on climate and atmospheric gas composition variability [50–52].

The model resolution is $5^\circ \times 4^\circ$ (longitude x latitude) with 39 vertical levels with variable spacing, up to 0.003 hPa at the model top [43]. The algorithm of the combined model accounts for the interaction between chemical and physical processes at each time step of the model. The chemical module accounts for 74 gas species interacting in 174 chemical reactions, and 46 reactions of photodissociation. A detailed description of the model can be found in [43,44]. For the numerical experiments we used the following datasets: the emissions of the ozone depleting substances were taken from [6]; the change in greenhouse gas emissions was parameterized according to the RCP4.5 scenario from the CMIP5 project; the data on extraterrestrial solar flux variability were taken according to [53], stratospheric aerosol concentrations followed [54]. The simulations were made using different datasets on SST/SIC (Sea Surface Temperature and Sea Ice Coverage, respectively): using the Met Office (as a default variant) [55], the ERA-Interim [42], and the CCM SOCOL (modeling tools for studies of SOlar Climate Ozone Links) datasets [56]. For evaluating *CMFuv*, a spectral correction was applied to the *CMF* data according to Equation (6).

2.2. The Description of ERA-Interim Reanalysis Dataset

The ERA-Interim reanalysis dataset [42] produced by the European Centre for Medium-Range Weather Forecasts (ECMWF) is a well-known reanalysis dataset available from 1979, continuously updated in real time. The ERA-Interim data-assimilation system utilizes a four-dimensional variational data-assimilation scheme (4D-Var) and a variational bias correction scheme (VarBC) for satellite radiances, that automatically detects and corrects for observation biases [57–60].

The ECMWF ERA-Interim reanalysis dataset, along with data on total ozone content, contains the data on solar radiation for cloudy and clear-sky conditions at the Earth's surface. For calculating *CMF*, we used the following ERA-Interim parameters: daily dose of "surface net solar radiation, clear sky" (NSR, clear sky) and "surface net solar radiation" (NSR), synoptic monthly means. It is easy to show that these parameters can be used for estimating *CMF* using the equation:

$$CMF_{i,j} = \frac{\mathrm{NSR}_{ij}}{\mathrm{NSR}_{\mathrm{clear\ sky}\ ij}}, \tag{7}$$

the UV correction was made using the same approach by Equation (6), which has been applied to the *CMF* data obtained from model simulations.

Validation of the ECMWF ERA-Interim ozone dataset has been made for the 1989–2008 period by comparisons with independent ground-based ozone observations [57] and satellite data [58]. In addition to ground-based independent observations, ozone profiles from ozonesondes retrieved from the World Ozone and Ultraviolet Radiation Data Centre (WOUDC) archive were used to validate the ozone vertical distribution. The residuals between the ERA-Interim and the ground-based Dobson total ozone measurements were within ±10% at high latitudes, and within ±5% over other regions with the absence of temporal bias both in the stratosphere and in the troposphere [57]. The validation of ERA-Interim relative to satellite data included total ozone data from TOMS and OMI satellite instruments as well as SBUV and SBUV/2 and ozone profile retrievals from other satellite measurements [58,59]. The ERA-Interim total column ozone uncertainty is typically ±5 DU (about ±2%). As a result, in the latest Scientific Assessment of Ozone Depletion: 2018 [8] on page 3.8 it was stated that "recent reanalysis datasets (Dee et al., 2011; Dragani 2011; and Wargan et al., 2017) have been shown to produce a realistic representation of total ozone".

ERA-Interim global solar irradiances at the ground have been validated with ground-based and satellite data [61–63]. According to a comparison of data from 674 sites, Era-Interim and ground-based measurements were highly correlated (R^2 = 0.97) [61]. Biases relative to observations were smaller compared to other reanalysis data evaluated in this study. According to Table 2 from [61], the mean bias relative to solar radiation measurements from all stations was 11.25 W m^{-2}, and the RMSE (Root-Mean-Square Error) was 27.7 W m^{-2}, while for the best radiation measurements at BSRN (Baseline Surface Radiation Network) sites the mean bias was 4.15 W m^{-2} and the RMSE was 19.6 W m^{-2}. A similarly small bias of 5.2 W m^{-2} was obtained over central Europe [62], while for the French radiation network the mean difference of 9.7 Wm^{-2} with $R^2 = 0.97$ was estimated on a daily basis for the 1995–2006 period [63].

Therefore, the estimates of total ozone and solar radiation in the ERA-Interim dataset can be characterized as realistic and suitable for our study.

2.3. Description of the Combined TOMS/OMI Satellite Dataset

The TOMS and OMI satellite datasets were used for comparisons with the *Eery* retrievals from the INM-RSHU CCM and the ERA-Interim reanalysis. For characterizing *Eery* variability, we applied daily erythemal doses (J m^{-2}) from the combined satellite dataset. Nimbus7/TOMS and EarthProbe/TOMS Level 3 data were downloaded from https://disc.gsfc.nasa.gov (date of access 01.02.2017) with a spatial resolution of 1° × 1.25° [64,65]. Level 3 AURA/OMI Erythemal Dose Daily product archives were downloaded from https://giovanni.sci.gsfc.nasa.gov (date of access 16.02.2017) with a spatial resolution of 1° × 1° [66]. The data were re-gridded to have similar spatial resolution. In addition, we made an updated correction on absorbing aerosol following the methodology described in [17,20,21]. For this purpose, we applied the new aerosol climatology Macv2 for 2005 [41]. The absorbing aerosol optical thickness (τ_{abs}) at λ = 320 nm was simulated using the following equation:

$$\tau_{abs} = \tau_{ext}(1-\omega) \quad (8)$$

where τ_{ext} is the extinction aerosol optical thickness; ω is the single scattering albedo.

According to [20,21,67], we evaluated aerosol correction factor *CF*, using the following equation:

$$CF = \frac{1}{1+K\tau_{abs}} \quad (9)$$

where the slope *K* weakly depends on solar elevation and aerosol type, and for $h > 30°$ with typical values of $\tau_{abs} < 0.1$ an average value of $K = 3$ can be applied with an error of less than 5% [20].

We should note that in OMI UV retrievals the absorbing aerosol correction had been already applied using the old Macv1 aerosol dataset [20,68]. Hence, for the application of the new Macv2 climatology correction we first removed the previous Macv1 correction from the dataset. For the TOMS dataset the correction using the new aerosol Macv2 dataset has been applied directly to the standard *Eery* retrievals. We would like to emphasize that both the previous and the new Macv2 corrections do not take into account changes in aerosols over time. The monthly mean τ_{abs} distribution with $1° \times 1°$ grid over the 1980–2015 period have been used as input data for the *CF* estimation. Figure 1 presents relative difference (in %) between the old (Macv1) and the new (Macv2) *CF* aerosol correction factor.

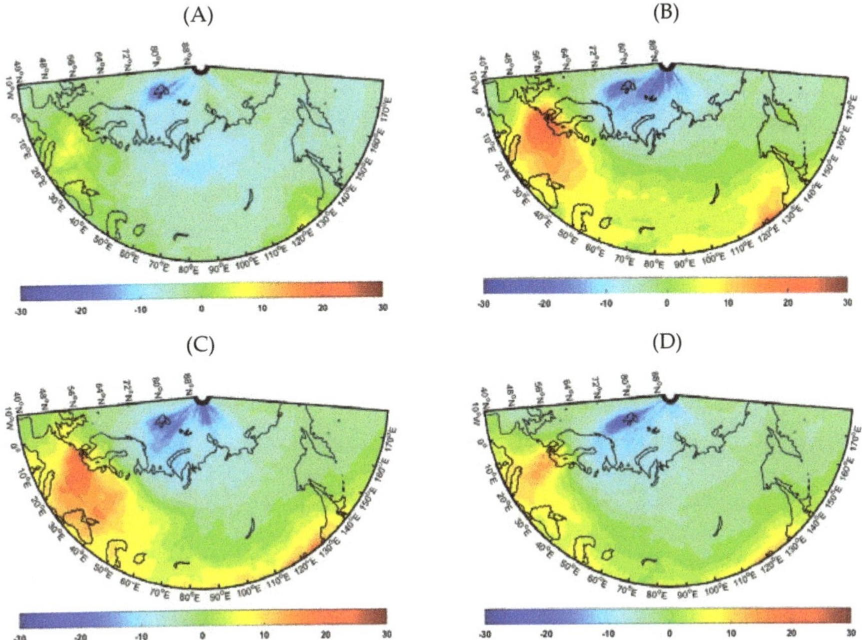

Figure 1. Relative difference (in %) between the new (Macv2) and the old (Macv1) aerosol correction factors, *CF* for the central months of the seasons: January (**A**), April (**B**), July (**C**), October (**D**).

Due to smaller absorbing aerosol optical depth over Europe and China in the new aerosol climatology we have relatively higher *CF*'s of up to 10–20% compared with the old Macv1 correction, especially in April and July. The opposite tendency is observed over the Arctic regions due to the increase in τ_{abs} in the new climatology, which resulted in 20–30% *CF* decrease almost for all seasons, except January. In other areas there is a slight change in *CF* of about ±5%. Due to changes in absorbing aerosol, there are pronounced differences in UV indices of more than 1 over several regions of Europe, Central Asia and China in July (Figure 2).

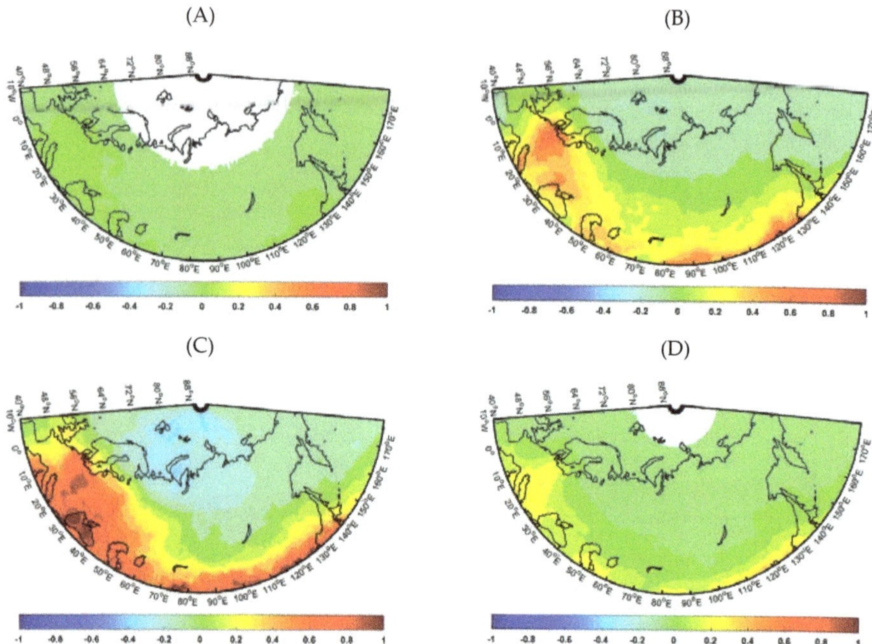

Figure 2. Absolute difference between the UV indices in 2005 with the application of the updated Macv2 *CF* factor minus UV indices with the standard OMI *CF* factor correction [68] for central months of the seasons: January (**A**), April (**B**), July (**C**), October (**D**). UV index is a widely used unitless quantity, defined by multiplying erythemal irradiance by 40 [1].

In addition, for characterizing the variations in *Eery* due to total ozone changes, the extended observational assimilated database described in [69] has been applied. The assimilated database combines satellite-based ozone measurements from TOMS, GOME, SBUV, and OMI satellite ozone instruments with their corrections from ground-based measurements. Since the development of this database was mainly based on TOMS and OMI total ozone retrievals, we refer to it in this study as the TOMS/OMI ozone dataset. This assimilated database has been also actively used in [8] for ozone trend studies.

We also applied the *CMFuv* climatology obtained previously according to the TOMS Reflectivity measurements (1979–2002) with additional surface albedo correction as described in [14,70]. This dataset has been used for the comparisons with the *CMFuv* climatology evaluated from the INM-RSHU CCM and ERA-Interim datasets. We did not use satellite TOMS/OMI *CMFuv* retrievals for characterizing year-to-year variability, since they have large uncertainty at high latitudes due to high snow surface albedo and low solar elevations [21]. We should mention that the results, obtained using satellite data in other publications [17–19], have been also restricted to 50°–55° latitude to avoid significant uncertainties.

2.4. The UV Resources Simulations

For characterizing the changes in erythemal UV radiation over the 1979–2015 period and for qualifying the importance of such changes for human health we used the UV resources approach, which was described in details in [14]. As stated above, the UV resources characterize UV radiation from the point of beneficial (vitamin D production) or adverse (sunburn) effects on human health in the categories of UV deficiency, UV optimum and UV excess with the additional account for skin types and open body fraction. The main principle of the approach is in the application of two thresholds. The first threshold is defined by the Minimum Erythemal Dose, *MED*, which depends on skin type

(k) according to the Fitzpatrick classification [71]. In this study we consider four skin types, which are typical for Northern Eurasia area. The *MED* for the most UV susceptible skin type 1 is 210 J m^{-2}, for skin type 2–250 J m^{-2}, for skin 3–350 J m^{-2}, and for skin type 4–450 J m^{-2} [71]. For higher UV levels which exceed *MED*, in [14] we proposed to define several subclasses of *UV excess*. They are attributed to the thresholds depending on the WHO (World Health Organization) UV index (*UVI*) categories [72]: *moderate UV excess* category, *high UV excess* category, *very high UV excess* category, and *extremely high UV* excess category, which are respectively related to moderate, high, very high, and extremely high hourly *UVI* (see the details in [14]).

The second threshold is the minimum vitamin D dose production *MvitDD*. A detailed description of the method of its evaluation can be found in [14]. It has been estimated with account for a daily dietary vitamin-D intake of 1000 IU [73] and using an equivalent of vitamin D production of one *MED* of 10,000 IU [74]. Hence, taking into consideration an open body fraction S, a relationship between the two thresholds can be written as follows:

$$MvitDD_k = 0.1\ MED_k/S \tag{10}$$

These two thresholds were used for defining different classes of UV resources including the UV deficiency, when no vitamin D is possible to obtain, UV optimum, when it is possible to obtain vitamin D and not possible to get sunburn (erythema), and UV excess, when erythema is obtained. We should note that S was set to 0.25 in this study. This simplified approach provides an opportunity in a first approximation to quantify the UV resources and their spatial and temporal changes.

For evaluating UV resources, absolute *Eery* estimates are needed. For this purpose, we calculated the hourly *Eery* averages centered at noon with 3-min resolution, using the 8-stream discrete ordinate radiative transfer method in the TUV model [48] over the territory of 10.5° W–179.5° E, 40° N–80° N with 1° × 1° increments for year 2000 for clear-sky conditions. In these simulations we took into account total ozone as well as aerosol optical thickness and surface albedo for central days of each month of 2000. To simulate *Eery* in cloudy conditions, we applied the CMF_{UV} values. Since *Eery* anomalies in Equation (1) were also simulated relative to 2000, we could easily reconstruct absolute *Eery* changes over the whole period.

Previously, the climatology of UV resources over Northern Eurasia has been obtained in [14]. In this paper we use this approach for evaluating possible changes in spatial distribution of the different UV resources classes since 1979 for different seasons and for various skin types due to ozone and cloud variations.

3. Results

3.1. Climatologies of Total Ozone and Cloud Modification Factor over Northern Eurasia

Since total ozone is one of the main factors affecting erythemal UV radiation, it is necessary to calculate both the ozone climatology and its temporal variation with low uncertainty. The comparisons of different ozone climatologies over the 1979–2015 period obtained from INM-RSHU CCM, ERA-Interim Reanalysis and Satellite TOMS/OMI datasets are shown in Figure 3 and Table 1, where the main statistics of total ozone column are presented. The CCM adequately reproduces the absolute values of total ozone amount: the differences between monthly mean CCM total ozone values do not exceed 7% compared with TOMS/OMI ozone retrievals with mean annual difference of about 1% (see Table 1).

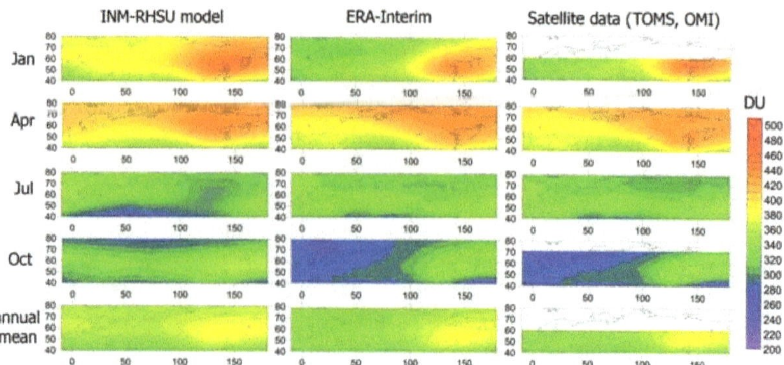

Figure 3. Climatology of the total ozone content (in Dobson units) over 1979–2015 according to the INM-RSHU CCM, ERA-Interim reanalysis, and TOMS/OMI satellite data. *X*-axis—is longitude, *Y*-axis—is latitude.

Table 1. Main statistics of total ozone column X (DU) and cloud modification factor *CMFuv* over Northern Eurasia for the 1979–2015 period *.

	January	April	July	October	Year
			Total Ozone Column, X, DU Mean ± confidence interval at $p = 95\%$		
INM-RSHU CCM	390 ± 4	384 ± 4	306 ± 2	320 ± 2	350 ± 3
ERA-Interim	365 ± 1	387 ± 1	326 ± 0.3	302 ± 1	346 ± 1
TOMS/OMI	370 ± 1	386 ± 1	326 ± 0.4	302 ± 1	346 ± 1
			Median		
INM-RSHU CCM	385	385	304	320	352
ERA-Interim	350	381	327	293	336
TOMS/OMI	354	380	327	293	336
			Standard Deviation		
INM-RSHU CCM	32	28	17	17	21
ERA-Interim	40	25	9	18	20
TOMS/OMI	38	26	11	19	20
			Case Number (Number of Pixels)		
INM-RSHU CCM	234	234	234	234	234
ERA-Interim	3990	3990	3990	3990	3990
TOMS/OMI	3192	3192	3192	3192	3192
			Cloud Modification Factor, *CMFuv* **Mean**		
INM-RSHU CCM	0.83 ± 0.01	0.80 ± 0.01	0.79 ± 0.01	0.80 ± 0.01	0.80 ± 0.01
ERA-Interim	0.77 ± 0.003	0.82 ± 0.002	0.79 ± 0.003	0.78 ± 0.003	0.78 ± 0.002
TOMS (1979–2002)	0.64 ± 0.003	0.73 ± 0.003	0.73 ± 0.003	0.69 ± 0.003	0.69 ± 0.002
			Median		
INM-RSHU CCM	0.82	0.80	0.78	0.79	0.81
ERA-Interim	0.75	0.82	0.78	0.77	0.77
TOMS (1979–2002)	0.64	0.73	0.73	0.67	0.68
			Standard Deviation		
INM-RSHU CCM	0.07	0.06	0.11	0.05	0.07
ERA-Interim	0.10	0.06	0.08	0.09	0.07
TOMS (1979–2002)	0.10	0.08	0.10	0.09	0.07
			Case Number (Number of Pixels)		
INM-RSHU CCM	234	234	234	234	234
ERA-Interim	4158	4158	4158	4158	4158
TOMS (1979–2002)	4158	4158	4158	4158	4158

* For comparison purpose the calculations were made only over the same areas.

There is a good agreement between ERA-Interim and TOMS/OMI mean total ozone datasets with mean differences of 1% or less. There is also a satisfactory agreement in the representation of ozone spatial distribution in different seasons, except October, when a dipole structure of ozone maximum has been obtained with the centers over the Western and Eastern coast of Northern Eurasia, while according to the satellite TOMS/OMI dataset we see only one ozone maximum over Northeastern Asia. There is a high correlation between TOMS/OMI and ERA-Interim Reanalysis datasets since in ERA-Interim ozone is both modeled and assimilated from SBUV, OMI, TOMS, GOME, and SCIAMACHY datasets [57,58].

The effect of cloud on UV radiation is a combination of the effects of cloud optical properties and spatial cloud characteristics (cloud amount). As shown above, cloud modification factor (*CMFuv*) captures both effects simultaneously. Figure 4 shows the spatial distribution of *CMFuv* according to INM-RSHU CCM, ERA_INTERIM, and TOMS/OMI satellite datasets over the 1979–2015 period. Main statistics of *CMFuv* are shown in Table 1. Noticeable variability in *CMFuv* reflects significant changes of different synoptic processes over Northern Eurasia. In January in midlatitudes cloud modification factors over Europe are much smaller due to the enhanced cyclonic activity (*CMFuv* = 0.3–0.4) and over the central part of the continent the *CMFuv* can reach 0.8–0.9 due to the influence of the Siberian anticyclone, which is characterized mainly by cloudless conditions. In July over the southern European regions one can see high *CMFuv* > 0.9 due to the influence of Azores anticyclone. All datasets are in reasonably agreement, except satellite dataset over Arctic area, where in April and July there is a significant drop in *CMFuv* possibly due to the problems with dividing cloud/albedo effects in the satellite algorithm, that results in unrealistically low *CMFuv* values.

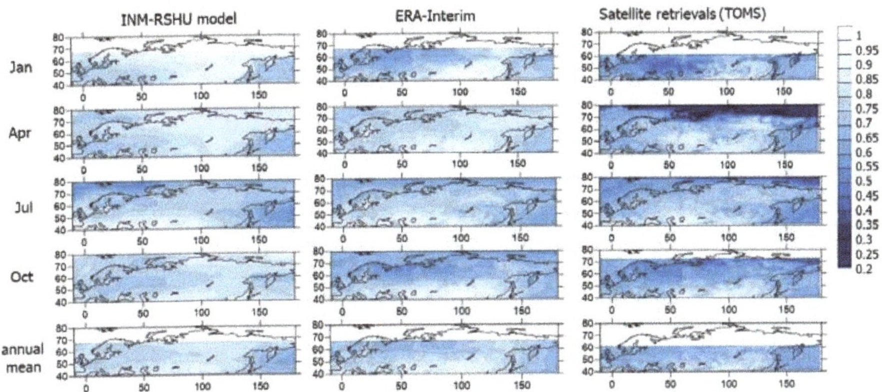

Figure 4. Climatology of the cloud modification factor (*CMFuv*) according to different datasets over the 1979–2015 period. Note, that the satellite *CMFuv* climatology has been obtained according to the TOMS data for the 1979–2002 period according to [14]. *X*-axis—is longitude, *Y*-axis—is latitude.

There is a good agreement between all three datasets in July, while in January we observe the *CMFuv* overestimation over the continental areas of Northern Eurasia in the INM–RSHU CCM.

Table 1 also demonstrates the low TOMS/OMI *CMFuv* values compared with the INM–RSHU CCM and ERA-Interim *CMFuv* retrievals, especially in January, in high snow albedo conditions, when the negative bias can reach 20–30%. On average, there is a satisfactory agreement between the mean *CMFuv* values in the INM-RSHU CCM and ERA-Interim datasets with the difference not exceeding 2.5%, except January, when it increased up to 8% (Table 1). We also should mention that previous comparisons with the direct *CMFuv* evaluation from the Moscow State University Meteorological Observatory dataset also demonstrate a satisfactory agreement with the INM–RSHU CCM *CMFuv* values with the exception of winter months [28].

3.2. *Changes in Eery Daily Doses due to Ozone and Cloudiness*

3.2.1. Eery due to Total Ozone Variations

There are several factors affecting ozone content in the atmosphere and, hence, noticeably change *Eery* at ground. Due to the Montreal Protocol with its Amendments and Adjustments the emissions of ozone depleting substances (ODS) have been restricted, however, due to significant life time of most of them, their concentrations in the atmosphere are getting lower only in recent years [6,8,10,38]. Along with the anthropogenic effects, the observed variability of some natural factors also could be important in assessing the total variability of ozone. We performed several numerical experiments for evaluating and comparing the role of the main ozone drivers in *Eery* variability over the 1979–2015 period. The detailed description of the numerical experiments is shown in Table 2.

Table 2. The description of the numerical INM-RSHU CCM experiments shown in Figures 5 and 6.

Names	Factors				All Factors Together
	Anthropogenic Effect	**Solar Activity**	**Stratospheric Aerosol**	**SST/SIC**	
Description	ODS changes were accounted over the 1979–2015 period according to [6]. All other parameters were set for 1979.	The changes in extraterrestrial solar fluxes were accounted over the 1979–2015 period according to [53]. All other parameters were set for 1979.	The changes in stratospheric aerosol were accounted over the 1979–2015 period according to [54]. All other parameters were set for 1979.	The changes in SST/SIC were accounted over the 1979–2015 period according to [55]. All other parameters were set for 1979.	The changes of all factors were accounted over the 1979–2015 period.

In order to estimate the sensitivity of different factors we simulated the difference in *Eery* between the periods of 2000–2015 and 1979–1999 due to the changes in emissions of ODS, the changes in stratospheric aerosol, in SST/SIC, and solar activity separately and due to their combined effects according to the numerical experiment, when all factors are taken into account simultaneously. The results are shown in Figure 5. Over this period there is a noticeable enhancement in *Eery* due to the ozone loss related to the ODS increase. The largest ozone *Eery* growth of more than 4% is observed over polar regions, which is in accordance to other model simulations [7].

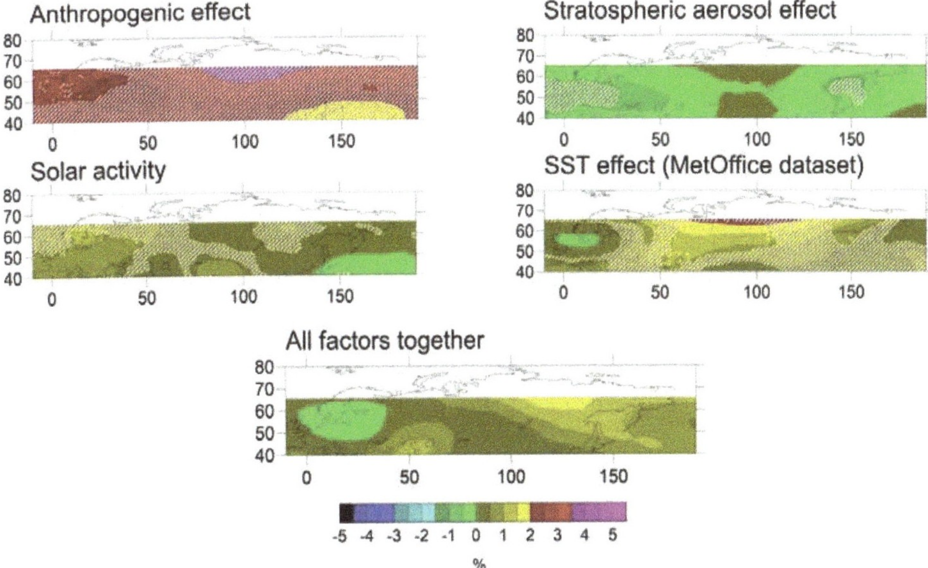

Figure 5. Difference in the average of annual Eery anomalies for the period 2000–2015 and the period 1979–1999, due to changes in total ozone column according to effects of different anthropogenic and natural factors. INM-RSHU CCM simulations. Statistically significant difference at $p = 95\%$ are shown by white hatching. *X*-axis—is longitude, *Y*-axis—is latitude.

Solar activity provides the changes in solar radiation at wavelengths shorter than 242 nm, which influence the ozone photochemical processes. Due to a tendency of decreasing solar activity during the last decades, smaller ozone amount is generated providing small, but positive changes in *Eery* (<1%). The increase in concentration of stratospheric aerosol was mainly observed due to strong volcanic eruptions at the end of the 20th century—El-Chichon (1982) and Pinatubo (1991). They resulted in ozone loss initiated by the heterogeneous processes on the surface of sulphate volcanic aerosol particles in the stratosphere. The absence of intensive volcanic activity in the 21 century provided a small *Eery* decrease during the 2000–2015 period (less than 1%) compared with the 1979–1999.

The changes in SST/SIC have the strongest impact on ozone compared with other natural factors presumably due to their influence on the atmosphere dynamic processes. These changes provide increase in *Eery* up to 2% for annual means in 21th century (2000–2015), especially over the northern Arctic regions. The numerical experiment, which took into account for all factors influencing total ozone column, has revealed a positive change in *Eery* up to 1–2% over northern regions of Eurasia and some small negative effects (less 1%) at the southeastern Pacific Ocean region, which are not statistically significant.

Since the effects of sea surface temperature on ozone are important, we also analyzed the possible changes in *Eery* due to the application of the other SST/SIC datasets in the INM-RSHU CCM (the ERA-Interim [42], and the SOCOL SST/SIC datasets [56]). In these numerical experiments the total ozone column was simulated using the same datasets for ODS, stratospheric aerosol and solar activity as before, but the MetOffice SST/SIC dataset was replaced with ERA-Interim and SOCOL data.

The comparisons of the influence of different SST/SIC on variations in *Eery* have revealed small difference in annual *Eery* behavior, especially between the model runs with the Met Office and the SOCOL SST/SIC. All numerical experiments have a similar tendency of annual *Eery* growth of about 1–2% over the central part of Northern Siberia in 2000–2015 compared with 1979–1999, however, there are large differences in seasonal changes. We would like to emphasize that the influence of SST/SIC on variations in *Eery* should be studied further for understanding the physical mechanism of this phenomenon. We may assume that the changes in SST/SIC may influence the intensity of NAO/AO (North Atlantic Oscillation/Arctic Oscillation) and AAO (Antarctic Oscillation) indices, which is small in zonal averages [75], but according to [8], their variations can explain much of the variability in ozone over local areas [76,77].

In addition, we analyzed variations in *Eery* due to the anthropogenic and natural ozone drivers during the 1979–2015 period using linear trend analysis (Figure 6). Similar to Figure 5, statistically significant positive linear *Eery* trends at p = 95% of up to 3%/decade have been obtained according to the numerical experiment with the ODS change (marked as "anthropogenic effect"). The sign of the *Eery* trends due to other natural factors are also similar to *Eery* changes shown in Figure 5.

The numerical experiment, which accounted for all factors, was tested against different datasets. Figure 7 shows the difference in the average of *Eery* anomalies for the period 2000–2015 and the period 1979–1999 due to ozone factor using the *Eery* retrievals according to the INM-RSHU CCM, the ERA-Interim datasets, and TOMS/OMI satellite data. There is a satisfactory agreement between annual results with a 1–2% *Eery* increase due to lower ozone in the 2000–2015 period compared to the 1979–1999 period. The annual Eery changes were not statistically significant, except the *Eery* increase over Eastern Europe and Northeastern Asia according to the ERA-Interim *Eery* retrievals (up to 3%). The discrepancy in *Eery* changes due to ozone in different datasets is much higher for the seasons. While the results for the ERA-Interim and TOMS/OMI satellite datasets are in reasonable agreement for April and July, the agreement for January and October is poor. The *Eery* increase obtained in the INM-RSHU CCM in April is mainly in agreement in sign with ERA-Interim and satellite data with more than 5% increase due to lower ozone content in 2000–2015, however, the spatial pattern of changes in *Eery* does not match other datasets. In July and October large territories are characterized by the decrease in model *Eery* due to more rapid recovery of total ozone compared with that in the ERA-Interim and the TOMS/OMI datasets. Similar tendency in *Eery* change has been also obtained in the detailed analysis of trends in *Eery* over Moscow [28].

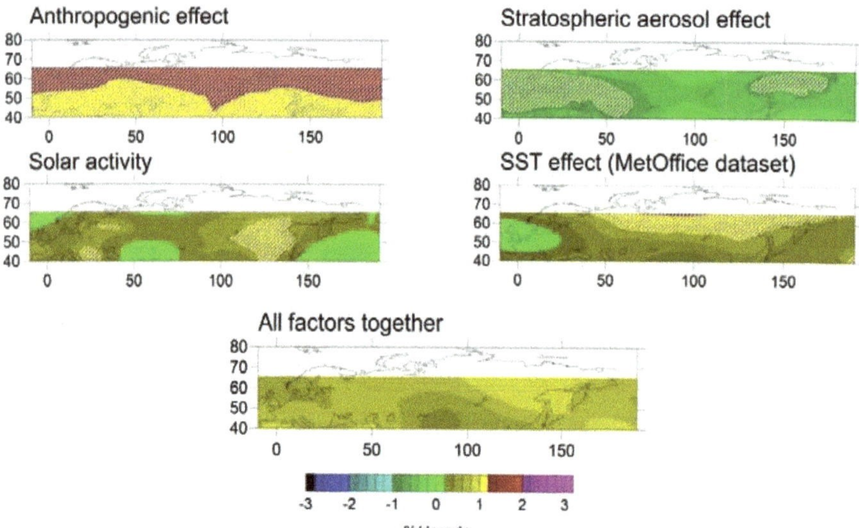

Figure 6. Decadal trends in annually averaged *Eery* anomalies (%/decade) due to changes in total ozone column according to the effects of anthropogenic (ODS) and different natural factors. INM-RSHU CCM simulations, 1979–2015. Statistically significant trends at 95% are shown by white hatching. *X*-axis—is longitude, *Y*-axis—is latitude.

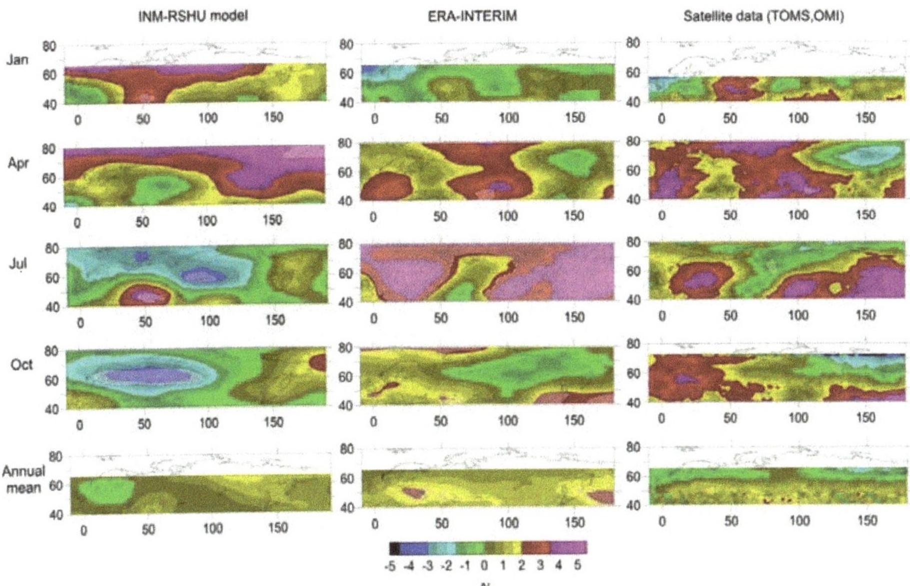

Figure 7. Difference in the average of *Eery* anomalies for the period 2000–2015 and the period 1979–1999 due to total ozone changes calculated from the different datasets. Statistically significant differences at 95% are shown by white hatching. Note that the left-bottom panel is identical to the bottom panel of Figure 5. *X*-axis—is longitude, *Y*-axis—is latitude.

In addition, Figure 8 presents time series of *Eery* anomalies at 48° N for longitudes between 0° E and 150° E based on the INM-RSHU CCM, ERA-Interim, and TOMS/OMI satellite datasets. The results from the three datasets agree within a few percent and suggest that *Eery* has increased between 1979 and 1995. In contrast, *Eery* has not changed perceivably between 2000 and 2015. These results corroborate the conclusion from Figure 8 that *Eery* was lower during 1979–1999 compared to 2000–2015.

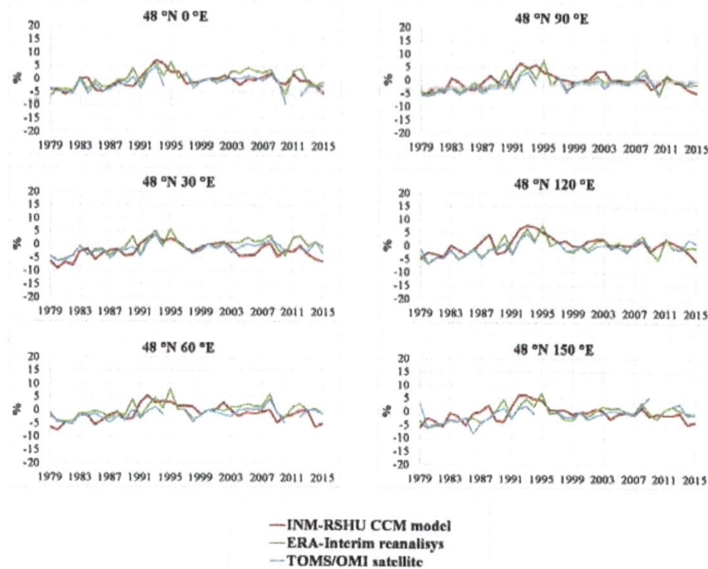

Figure 8. The time series of the *Eery* relative anomalies against 2000 at 48° N from 0° E to 150° E according to the INM-RSHU CCM, ERA-Interim, and TOMS/OMI satellite datasets.

Figure 9 shows the spatial distribution of the decadal trend in *Eery* caused by changes in total ozone over the period 1979–2015, calculated from the INM-RSHU CCM, ERA-Interim, and TOMS/OMI satellite datasets. For annual means (last row in Figure 9), trends computed from the three datasets are generally consistent and range between −0.5% and 2% per decade. However, similar to the results shown in Figure 7, the discrepancy is much larger for the seasonal datasets. In April a pronounced increase in *Eery* (up to 3% per decade) is observed according to all datasets, which is statistically significant over vast areas. Results for the ERA-Interim and TOMS/OMI satellite datasets agree again reasonably well, while results from the INM-RSHU CCM model in most cases do not capture the spatial patterns of the observations.

3.2.2. *Eery* due to Cloud Variations

We analyzed *CMFuv* variations retrieved from the INM-RSHU CCM numerical experiments as well as from the ERA-Interim dataset. The seasonal and annual differences in the *CMFuv* values between the 2000–2015 and 1979–1999 periods are shown in Figure 10. In general, the *CMFuv* differences vary between these periods within ±5%. However, in April over Central and Eastern Europe and in July over the northern Atlantic, the European territory of Russia, Central Siberia and Northeastern Asia the difference is higher than 5% and sometimes exceeds 10% according to the ERA-Interim dataset. These estimates are in a good agreement with the *CMFuv* trends over Moscow [78]. Model simulations provide much smaller changes in *CMFuv*. Similar tendencies are observed only over the Atlantic and Northeastern Asia, but at much smaller level. In contrast, in July over the Northern Arctic region there is a statistically significant *CMFuv* decrease, which has been obtained both according to the INM-RSHU CCM and the ERA-Interim datasets. This may happen due to increasing temperature in this northern area as a result of global warming and intensification of the cyclonic processes and, hence, increase in water vapor content. We should emphasize that these results are also in agreement with the independent model simulations [11]. In addition, in [79] it was mentioned that by the end of the 21 century "cloud cover will increase at high latitudes by up to 5% but will decrease at low latitudes (<~30°) by up to 3%". According to [80] a UV reduction of 10–15% is projected by 2100 due to increases in cloudiness over some northern high-latitude regions and over Antarctica. Hence, the tendency in increasing of cloud cover in summer Arctic conditions is successfully retrieved from INM-RSHU CCM. At the same time, the INM-RSHU CCM does not reproduce well positive changes in *CMFuv* over midlatitudes. For annual data, one can see even a prevailing small model *CMFuv* decrease compared with the *CMFuv* increase of up to 5% over vast midlatitude areas according to the ERA-Interim dataset.

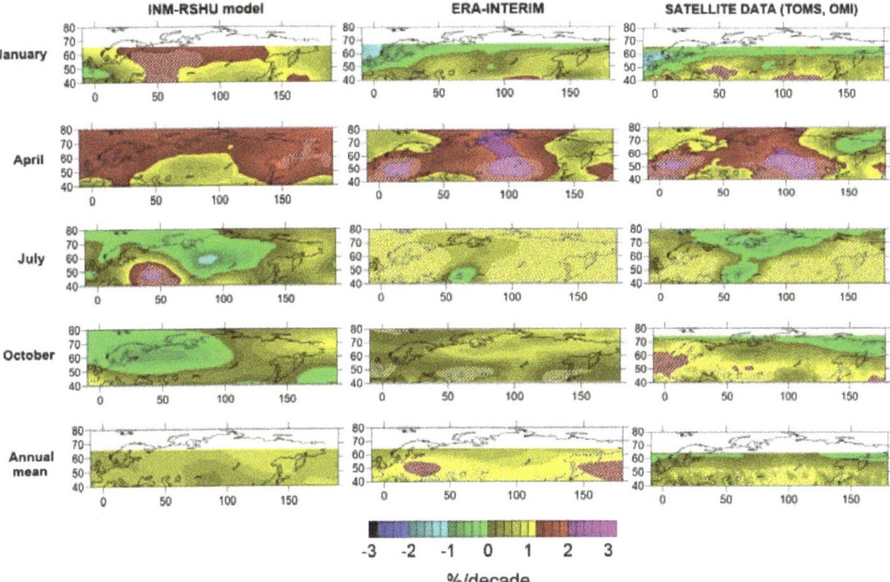

Figure 9. Spatial distribution of the decadal trend in *Eery* caused by changes in total ozone over the period 1979–2015, calculated from the INM-RSHU CCM, ERA-Interim, and TOMS/OMI satellite datasets. Statistically significant differences at 95% are shown by white hatching. Note that the left-bottom panel is identical to bottom panel of Figure 6. *X*-axis—is longitude, *Y*-axis—is latitude.

The additional linear trend analysis also demonstrates high *CMFuv* increase (up to +4–6% per decade) over several areas (East-European Plain, Northeastern Asia, and local areas in Siberia) according to the ERA-Interim dataset (Figure 11), while model positive linear decadal trends are much smaller. Over the Arctic basin negative decadal trends of about 2–4%/decade have been revealed from both model and ERA-Interim *CMFuv* retrievals in April and July. This is also in agreement with the results obtained in [11]. We should also note that the model does not reproduce positive trends in *CMFuv* over Central Europe, Central Siberia and Northeastern Asia in July, and over the Northeastern Arctic regions in October.

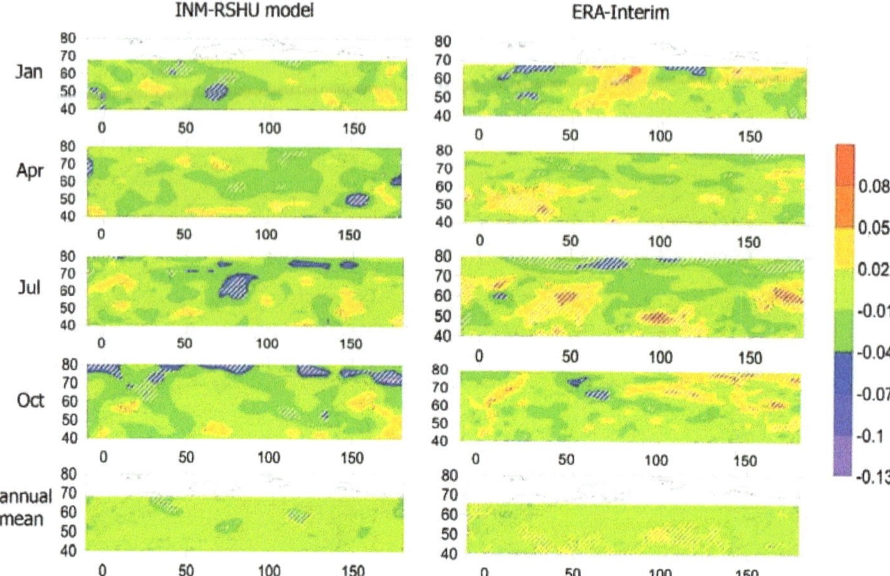

Figure 10. Difference in the average of *CMFuv* anomalies for the period 2000–2015 and the period 1979–1999 according to the INM-RSHU CCM simulations and the retrievals from the ERA-Interim dataset. Statistically significant differences at $p = 95\%$ are shown by white hatching. *X*-axis—is longitude, *Y*-axis—is latitude.

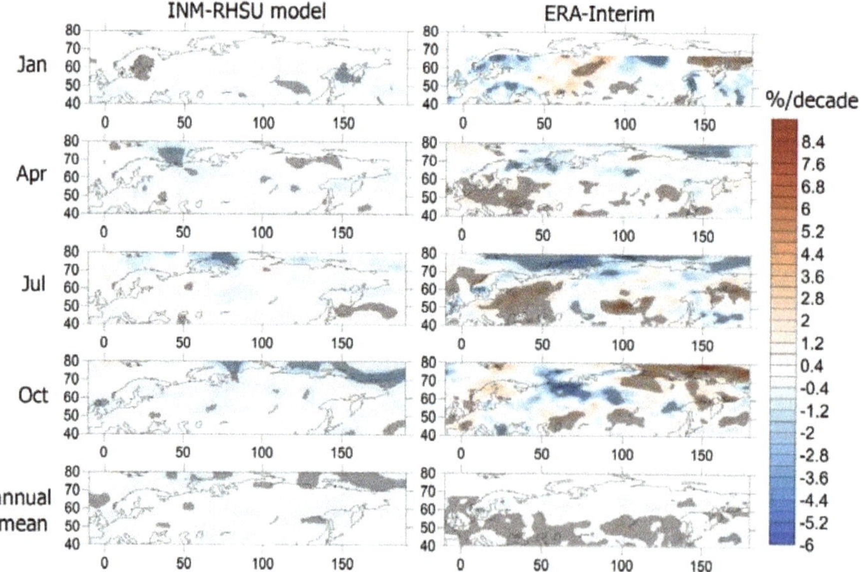

Figure 11. The decadal trends in *Eery* due to changes in cloudiness according to the INM-RSHU CCM simulations and the retrievals from the ERA-Interim dataset over the 1979–2015 period. Statistically significant trends at 95% level are shown by hatching. *X*-axis—is longitude, *Y*-axis—is latitude.

3.2.3. *Eery* Changes due to Joint Influence of Total Ozone and Cloud Variations

According to the method, described in Section 2 (Equations (1) to (6)), we simulated variations in *Eery* due to both factors: ozone and cloudiness. We applied this approach for both ERA-Interim and INM-RSHU CCM datasets. In addition, we made the comparisons with TOMS/OMI daily erythemal doses dataset with the new Macv2 aerosol correction. Figure 12 shows seasonal and annual differences in *Eery* anomalies between the 2000–2015 and 1979–1999 periods due to the combined effects of total ozone and cloudiness. For annual means (last row in Figure 12), results for the CCM model show increases by 0–3% in 2000–2015, while changes for the ERA-Interim dataset range between −2% and 5%, with few exceptions. In July the spatial changes in *Eery* from the Era-Interim dataset exhibit a quasi-wave spatial structure in midlatitudes with a significant increase of about 10–15% over Europe, Central Siberia, and Northeastern Asia. A similar structure, albeit less pronounced, is also apparent in satellite retrievals, and the CCM model. The agreement in the spatial patterns of the datasets is generally poor for other months, except areas over oceans. This disagreement may partly be caused by the interaction between clouds and high surface albedo from snow cover (e.g., see discussion in [67]). This interaction is particularly a problem over northern areas that are affected by snow cover from October through June.

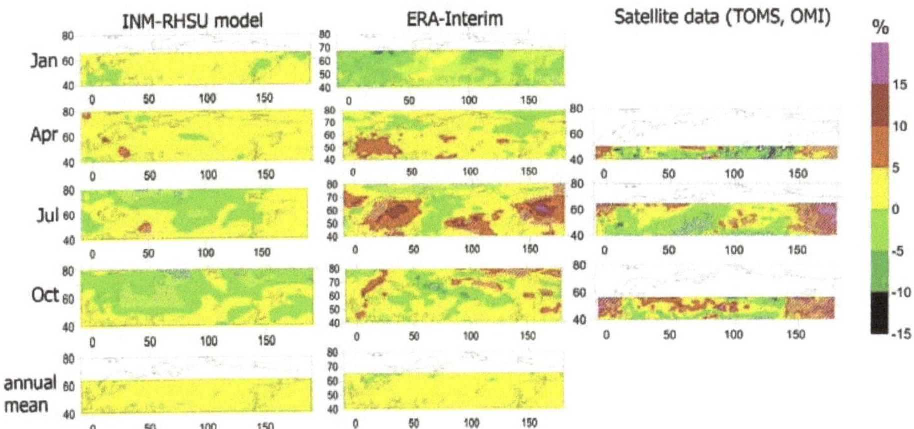

Figure 12. Seasonal and annual differences in *Eery* anomalies between the 2000–2015 and 1979–1999 periods due to combined effects of changes in total ozone column and cloudiness according to the INM-RSHU CCM model, the ERA-Interim dataset, and Eery retrievals from the TOMS/OMI satellite dataset with the additional Macv2 aerosol correction. Statistically significant difference at 95% are shown by hatching. Note, that we do not present UV retrievals from the TOMS/OMI dataset for January and for annual mean due to the problems in UV retrievals in conditions with high surface snow/ice albedo. *X*-axis—is longitude, *Y*-axis—is latitude.

Figure 13 presents *Eery* decadal trends due to the combined effect of total ozone and cloudiness according to the INM-RSHU CCM, ERA-Interim simulations, and *Eery* retrievals from the TOMS/OMI satellite dataset with the additional aerosol Macv2 correction. For annual means (last row in Figure 13), trends calculated from the CCM and ERA-Interim datasets generally agree, range between 0% and 3% per decade, and are statistically significant over large areas. The noticeable positive trends in *Eery* of up to 8% per decade in July, and up to 5% per decade in April are observed in the ERA-Interim dataset over several areas in Eastern Europe, Central Siberia, and Northeastern Asia. In July, similar trends in *Eery* of up to 8% per decade can be found according to the satellite data over the Pacific Ocean. We also see a better agreement in July, when the spatial quasi-wave structure of positive *Eery* trends in midlatitudes from the ERA-Interim is similar to that obtained from the satellite data. Spatial patterns of the CCM and ERA-Interim results agree qualitatively, but trends calculated with the CCM model are generally smaller.

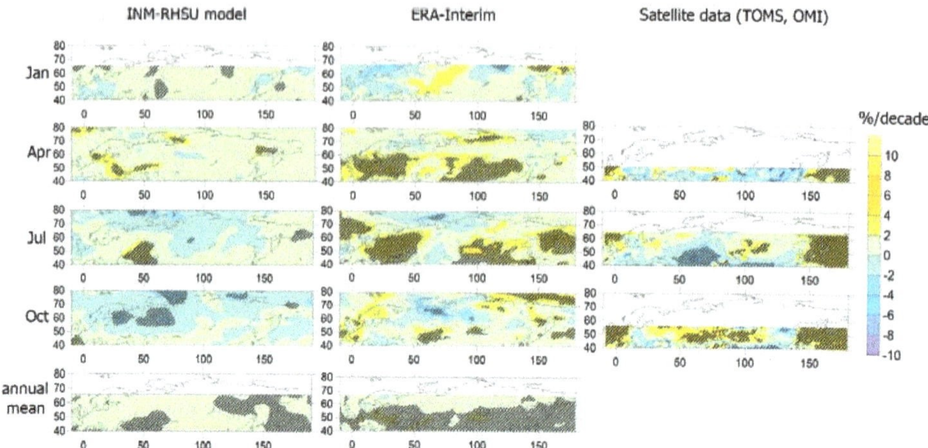

Figure 13. Spatial distribution of the decadal trend in *Eery* caused by the combined effect of total ozone and cloudiness calculated from the INM-RSHU CCM, ERA-Interim, and TOMS/OMI satellite datasets. Statistically significant differences at 95% are shown by hatching. Note, that we do not present UV retrievals from the TOMS/OMI dataset for January and for annual mean due to the problems in UV retrievals in conditions with high surface snow/ice albedo. *X*-axis—is longitude, *Y*-axis—is latitude.

3.3. *Changes in UV Resources due to the Changes in Ozone and Cloudiness over Northern Eurasia*

Using the method described in Section 2.4, we simulated the UV resources for 1979 and 2015 and their changes over this period according to the retrievals from the ERA-Interim data for different skin types. The *Eery* retrievals from the ERA-Interim dataset were chosen for the following reasons. For estimating *Eery*, we need the reliable products for both total ozone and cloud modification factor. A large territory of Northern Eurasia is located in high latitudes with small solar elevations and snow cover dominating not only in winter, but during spring and fall, when satellite *Eery* retrievals cannot be used. In Section 2.2 the Era-Interim total ozone and downward shortwave radiation products were shown to have a good quality [57–63]. Therefore, we simulated UV resources using the *Eery* retrievals from this dataset. To avoid local features in *Eery* variations in 1979 and 2015, we applied a linear regression model to characterize the *Eery* changes from 1979 to 2015.

Figure 14 presents the UV resources distribution over Northern Eurasia for 1979 and 2015 and their difference for skin types 1–4. According to [72], skin type 1 is most susceptible to erythema, while skin type 4 is characterized by small sun-sensitivity. We revealed that for the all skin types due to the reduction of ozone and clouds, there is a noticeable geographical shift of UV categories over large areas, especially in April and July. A noticeable reduction of the UV optimum area and its replacement by the moderate UV excess conditions is observed in April for skin types 1 and 2. There is also a shift from moderate to high UV excess conditions at lower latitudes for these skin types. However, for skin types 3 and 4 in April we see favorable changes from UV deficiency to the UV optimum conditions over polar regions, and adverse change of UV optimum to UV excess conditions at the south (see Figure 14). In July in central Arctic region there is, on the contrary, a change from UV excess to UV optimum conditions for skin types 2 and 3, due to an increase in cloudiness. In October favorable changes are observed for skin types 3 and 4 with UV optimum area shift towards the north, while for skin types 1 and 2 there is the replacement of UV optimum by moderate UV excess conditions at the south.

Skin type 1

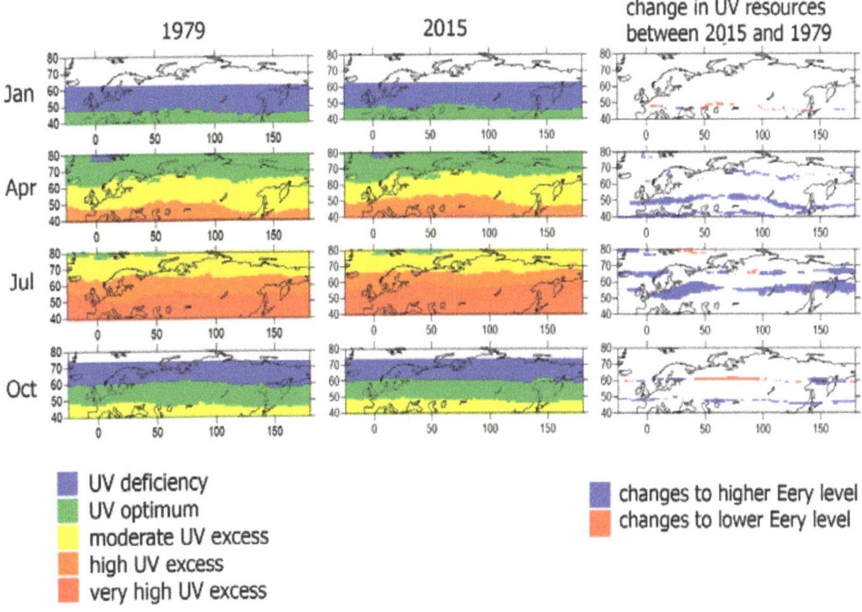

Skin type 2

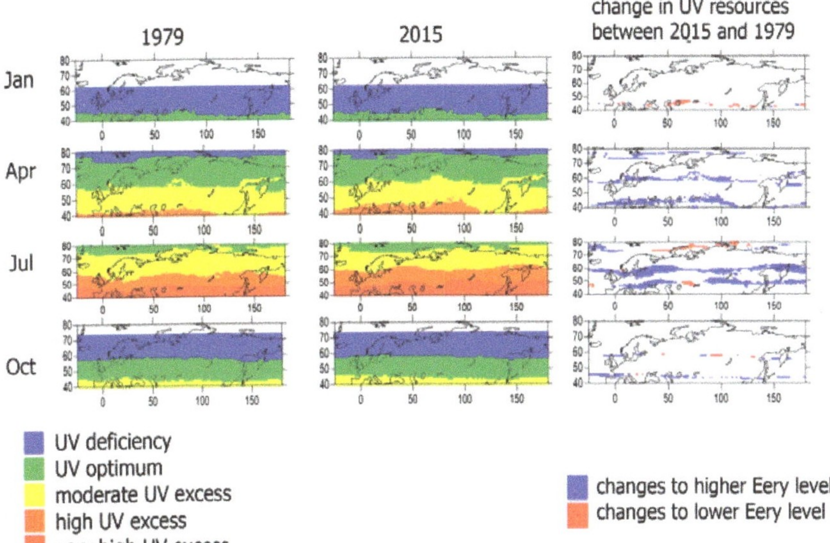

Figure 14. *Cont.*

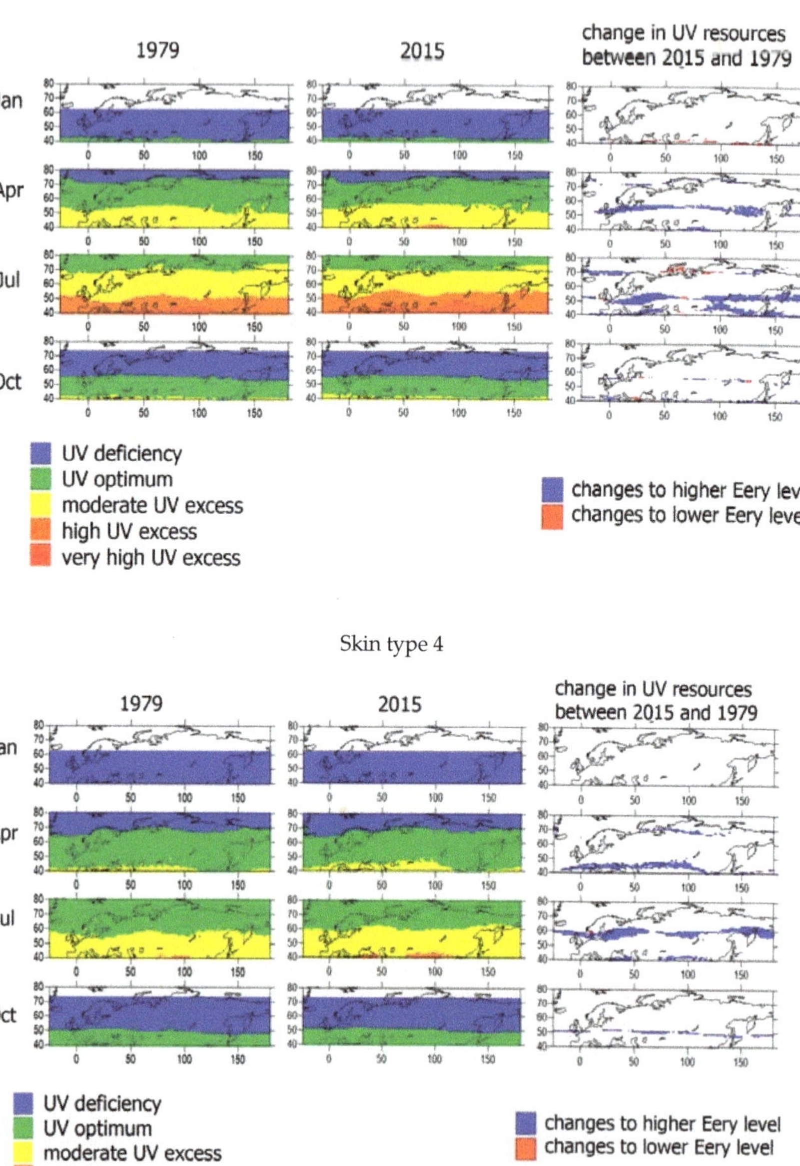

Figure 14. UV resources for 1979 and 2015 and changes in their distribution between 2005 and 1979 according to the retrievals from the ERA-Interim data for different skin types. *X*-axis—is longitude, *Y*-axis—is latitude.

4. Discussion

In this paper we evaluated the separate effects of total ozone and cloudiness and their combined effect on temporal variability of UV erythemal daily doses and UV resources according to the *Eery* retrievals from the ERA-Interim, the INM-RSHU CCM datasets, and TOMS/OMI satellite measurements with the Macv2 aerosol correction over Northern Eurasia for the 1979–2015 period.

The numerical experiments using INM-RSHU CCM model allowed us to estimate the role of main ozone drivers (ODS, stratospheric aerosol, sea surface temperature and ice coverage, solar activity) and to evaluate their impact on *Eery*. We showed that the most important factor affecting *Eery* is the concentration of ODS, the increase of which led to the significant ozone loss. Among natural factors, stratospheric aerosol and SST/SIC are the most important ones. The application of the Met Office and the SOCOL SST datasets in the INM-RSHU CCM provided the similar annual *Eery* increase in 2000–2015 over Central part of Northern Siberia, however, further studies should be conducted for understanding the physical mechanism of this phenomenon. On the whole, there is a positive change in the modeled *Eery* of up to 1–2% at the northern regions of Eurasia in 2000–2015 compared with the 1979–1999 period. Maximum positive linear *Eery* trends (up to 3% per decade) due to ozone loss were observed in April.

Linear *Eery* trends due to changes in cloudiness were much more pronounced and reached 4–8% per decade over some areas (East-European Plain, Northeastern Asia and local spots in Siberia) according to the ERA-Interim dataset. Both INM-RSHU CCM and ERA-Interim *Eery* retrievals have revealed negative *Eery* trends due to cloudiness of about 2–4% per decade in the central Arctic basin in April and July. This decrease in *CMFuv* during the last decades is connected with warming of the lower troposphere and large reduction of sea-ice-sheet, increasing evaporation and water vapor content [81]. Similar changes were obtained in [11], where it was mentioned that "the primary drivers of these changes are increasing concentrations of greenhouse gases (GHGs) and, for the southern hemisphere, the Antarctic ozone 'hole'". It should be emphasized that according to the model estimations *CMFuv* changes were much smaller than those obtained from ERA-Interim and satellite data. According to the ERA-Interim *Eery* retrievals, the joint ozone and cloud effect provided up to 6–8% increase in *Eery* per decade in July and April over several areas in Eastern Europe, Central Siberia, and Northeastern Asia. Annual *Eery* linear trends comprised about 2% per decade according to the ERA-Interim dataset and were statistically significant over large areas.

The analysis of the UV resources for 1979 and 2015 and their changes over this period showed a noticeable geographical shift of UV categories, especially in spring and summer over large areas due to the reduction of ozone and clouds. For skin types 1 and 2, which are most susceptible to sunburn, favorable UV optimum conditions were substituted by UV excess conditions over large areas. At the same time in April favorable UV optimum instead of UV deficiency conditions were observed for skin types 3 and 4 in the north but in the south the areas covered by the UV optimum have been replaced by the UV excess conditions. In some Arctic regions, in July, there was a change from UV excess to UV optimum for skin types 2 and 3, due to the increase in cloudiness.

In our assessment of UV resources, we did not take into account changes in the surface area of the skin that is exposed to sunlight or possible changes in the behavior of people, which could result from climate change and increasing temperatures. Instead, we focused on variations in UV resources due to the changes of the most important geophysical factors: total ozone and cloudiness.

5. Conclusions

We analyzed temporal variability of UV erythemal daily doses and UV resources due to ozone and cloudiness, according to different datasets, using *Eery* retrievals from the ERA-Interim reanalysis, INM-RSHU CCM, and the TOMS/OMI satellite measurements, with the updated aerosol correction over Northern Eurasia for the 1979–2015 period.

We showed that, according to all datasets for spring and summer clear-sky conditions, there was a pronounced trend in *Eery* of up to +3% per decade due to ozone loss, which was statistically significant over large areas in Northern Eurasia.

The INM-RSHU model experiments have confirmed that the largest effect on ozone and *Eery* stem from the largest impact of anthropogenic emissions of ozone-depleting substances. Additional factors include volcanic aerosol and the SST/SIC on ozone and, hence, on *Eery* changes. The utilization of the different SST/SIC datasets in the INM-RSHU CCM showed similar annual mean *Eery* increase over the polar region in Siberia.

The INM RSHU CCM, as many other CCM models (see, for example, discussion in [37]), did not reproduce an observed significant positive change in *CMFuv* during the last decades, which can reach up to 4–8% per decade, according to the ERA-Interim dataset over several areas in Northern Eurasia.

According to the *Eery* retrievals from the ERA-Interim dataset positive *Eery* trends reached 6–8% per decade over Eastern Europe, several regions in Siberia and Northeastern Asia due to the joint impact of ozone and cloudiness. In the central Arctic region, negative *Eery* trends were observed in summer due to *CMFuv* decrease, which is in agreement with findings in other studies [11].

The simulations of changes in UV resources from 1979 to 2015 have shown that there is a shift toward higher UV categories, especially in April and July, for skin types 1 and 2, which are most susceptible to sunburn. In April, this phenomenon was expressed in a noticeable reduction of the UV optimum area and its replacement by the moderate UV excess conditions. In addition, a shift from moderate to high UV excess conditions is observed. However, during summer months in the central Arctic region, there is a change from UV excess to UV optimum conditions for the skin types II and III due to a decrease in *CMFuv*.

This work is a part of an on-going project, within which we plan to perform additional numerical experiments with the INM-RSHU-CCM taking into the account temporal changes in aerosol and surface reflectivity as well as aerosol–cloud interaction, which possibly help in evaluating real *CMFuv* long-term changes over 1979–2015 period.

Author Contributions: The paper was initiated and written by N.E.C. All other authors had valuable contribution in visualization of the results and data analysis (A.S.P., E.Y.Z., and E.V.V.) and in providing the data of model experiments with the INM-RSHU CCM (S.P.S. and V.Y.G.); E.V.V. and E.Y.Z. helped with the technical preparation of the manuscript. All authors have read and agreed to the published version of the manuscript.

Funding: This research was partly supported in part by the Russian Foundation for Basic Research, grant no. #18-05-00700, and state budget program AAAA-A16-116032810086-4.

Acknowledgments: We would like to thank A. Arola for providing the aerosol dataset, which was used in the standard OMI *Eery* retrievals. We would also like to thank Bodeker Scientific, funded by the New Zealand Deep South National Science Challenge, for providing the combined NIWA-BS total column ozone database. In addition, the useful comments of the anonymous reviewers are highly acknowledged.

Conflicts of Interest: The authors declare no conflict of interest.

References

1. Robinson, S.A.; Wilson, S.R. *Environmental Effects of Ozone Depletion and Its Interactions with Climate Change: 2010 Assessment*; United Nations Environment Programme: Nairobi, Kenya, 2010; p. 328.
2. Environmental Effects Assessment Panel (EEAP). *Environmental Effects and Interactions of Stratospheric Ozone Depletion, UV Radiation, and Climate Change: 2018 Assessment Report*; United Nations Environment Programme (UNEP): Kenya, Nairobi, 2019; p. 390. Available online: https://ozone.unep.org/science/assessment/eeap (accessed on 1 October 2019).
3. Holick, M.F. Vitamin D: A millenium perspective. *J. Cell. Biochem.* **2003**, *88*, 296–307. [CrossRef]
4. Webb, A.R.; Engelsen, O. Ultraviolet exposure scenarios: Risks of erythema from recommendations on cutaneous vitamin D synthesis. In *Sunlight, Vitamin D and Skin Cancer*; Reichrath, J., Ed.; Advances in Experimental Medicine and Biology; Springer: New York, NY, USA, 2008; pp. 72–85.
5. Bornman, J.F.; Paul, N.; Shao, M. *Environmental Effects of Ozone Depletion and Its Interactions with Climate Change: 2014 Assessment*; United Nations Environment Programme: Kenya, Nairobi, 2014; p. 312.
6. World Meteorological Organization (WMO). *Scientific Assessment of Ozone Depletion: 2010*; Global Ozone Research and Monitoring Project—Report No. 52; World Meteorological Organization: Geneva, Switzerland, 2011; p. 516.

7. World Meteorological Organization (WMO). *Scientific Assessment of Ozone Depletion: 2014*; Global Ozone Research and Monitoring Project—Report No. 53; Meteorological Organization: Geneva, Switzerland, 2015; p. 416.
8. World Meteorological Organization (WMO). *Scientific Assessment of Ozone Depletion: 2018*; Global Ozone Research and Monitoring Project—Report No. 58; Meteorological Organization: Geneva, Switzerland, 2018; p. 588.
9. Bekki, S.; Bodeker, G.E.; Bais, A.F.; Butchart, N.; Eyring, V.; Fahey, D.W.; Kinnison, D.E.; Langematz, U.; Mayer, B.; Portmann, R.W.; et al. Future ozone and its impact on surface UV. In *Scientific Assessment of Ozone Depletion: 2010*; Global Ozone Research and Monitoring Project—Report 52; World Meteorological Organization: Geneva, Switzerland, 2011; Chapter 3; p. 516.
10. Bais, A.F.; Lubin, D.; Arola, A.; Bernhard, G.; Blumthaler, M.; Chubarova, N.; Erlick, C.; Gies, H.P.; Krotkov, N.; Mayer, B.; et al. Surface ultraviolet radiation: Past, present, and future. In *Scientific Assessment of Ozone Depletion: 2006, Global Ozone Research and Monitoring Project—Report No. 50*; World Meteorological Organization: Geneva, Switzerland, 2007; Chapter 7; p. 572.
11. Bais, A.F.; Bernhard, G.; McKenzie, R.L.; Aucamp, P.J.; Young, P.J.; Ilyas, M.; Jöckel, P.; Deushi, M. Ozone-climate interactions and effects on solar ultraviolet radiation. *Photochem. Photobiol. Sci.* **2019**, *18*, 602–640. [CrossRef] [PubMed]
12. Bodeker, G.; Burrowes, J.; Scott-Weekly, R.; Nichol, S.E.; McKenzie, R.L. A UV atlas for New Zealand. In Proceedings of the Workshop on UV Radiation and its Effects: An Update (2002), Auckland, New Zealand, 26–28 March 2002; pp. 26–28.
13. Fioletov, V.E.; Kimlin, M.G.; Krotkov, N.; McArthur, L.J.B.; Kerr, J.B.; Wardle, D.I.; Herman, J.R.; Meltzer, R.; Mathews, T.W.; Kaurola, J. UV index climatology over the United States and Canada from ground-based and satellite estimates. *J. Geophys. Res. Atmos.* **2004**, *109*. Available online: https://agupubs.onlinelibrary.wiley.com/doi/10.1029/2004JD004820 (accessed on 1 September 2019).
14. Chubarova, N.; Zhdanova, Y. Ultraviolet resources over Northern Eurasia. *J. Photochem. Photobiol. B Biol.* **2013**, *127*, 38–51. [CrossRef] [PubMed]
15. Lee-Taylor, J.; Madronich, S.; Fischer, C.; Mayer, B. A climatology of UV radiation, 1979–2000, 65S–65N. In *UV Radiation in Global Climate Change: Measurements, Modeling and Effects on Ecosystems*; Gao, W., Schmoldt, D.L., Slusser, J., Eds.; Springer: Dordrecht, The Netherlands; Tsinghua University Press: Beijing, China, 2009; Chapter 1; pp. 1–20.
16. Herman, J.R. Global increase in UV irradiance during the past 30 years (1979 to 2008) estimated from satellite data. *J. Geophys. Res. Atmos.* **2010**, *115*, D04203. [CrossRef]
17. Ialongo, I.; Arola, A.; Kujanpää, J.; Tamminen, J. Use of satellite erythemal UV products in analysing the global UV changes. *Atmos. Chem. Phys.* **2011**, *11*, 9649–9658. [CrossRef]
18. Herman, J.R.; Labow, G.; Hsu, N.C.; Larko, D. Changes in cloud and aerosol cover (1980–2006) from reflectivity time series using SeaWiFS, N7-TOMS, EP-TOMS, SBUV-2, and OMI radiance data. *J. Geophys. Res. Atmos.* **2009**, *114*. Available online: https://agupubs.onlinelibrary.wiley.com/doi/10.1029/2007JD009508 (accessed on 1 September 2019). [CrossRef]
19. Herman, J.; DeLand, M.T.; Huang, L.-K.; Labow, G.; Larko, D.; Lloyd, S.A.; Mao, J.; Qin, W.; Weaver, C. A net decrease in the Earth's cloud, aerosol, and surface 340 nm reflectivity during the past 33 yr (1979–2011). *Atmos. Chem. Phys.* **2013**, *13*, 8505–8524. [CrossRef]
20. Arola, A.; Kazadzis, S.; Lindfors, A.; Krotkov, N.; Kujanpää, J.; Tamminen, J.; Bais, A.; di Sarra, A.; Villaplana, J.M.; Brogniez, C.; et al. A new approach to correct for absorbing aerosols in OMI UV. *Geophys. Res. Lett.* **2009**, *36*. Available online: https://agupubs.onlinelibrary.wiley.com/doi/10.1029/2009GL041137 (accessed on 1 October 2019).
21. Krotkov, N.A.; Bhartia, P.K.; Herman, J.R.; Fioletov, V.; Kerr, J. Satellite estimation of spectral surface UV irradiance in the presence of tropospheric aerosols: 1. Cloud-free case. *J. Geophys. Res. Atmos.* **1998**, *103*, 8779–8793. [CrossRef]
22. Bernhard, G. Trends of solar ultraviolet irradiance at Barrow, Alaska, and the effect of measurement uncertainties on trend detection. *Atmos. Chem. Phys.* **2011**, *11*, 13029–13045. [CrossRef]
23. Cabrera, S.; Ipiña, A.; Damiani, A.; Cordero, R.R.; Piacentini, R.D. UV index values and trends in Santiago, Chile (33.5° S) based on ground and satellite data. *J. Photochem. Photobiol. B Biol.* **2012**, *115*, 73–84. [CrossRef]

24. Outer, P.N.; Slaper, H.; Kaurola, J.; Lindfors, A.; Kazantzidis, A.; Bais, A.F.; Feister, U.; Junk, J.; Janouch, M.; Josefsson, W. Reconstructing of erythemal ultraviolet radiation levels in Europe for the past 4 decades. *J. Geophys. Res. Atmos.* **2010**, *115*. Available online: https://agupubs.onlinelibrary.wiley.com/doi/10.1029/2009JD012827 (accessed on 15 September 2019).
25. Fitzka, M.; Simic, S.; Hadzimustafic, J. Trends in spectral UV radiation from long-term measurements at Hoher Sonnblick, Austria. *Theor. Appl. Climatol.* **2012**, *110*, 585–593. [CrossRef]
26. Lemus-Deschamps, L.; Makin, J.K. Fifty years of changes in UV Index and implications for skin cancer in Australia. *Int. J. Biometeorol.* **2012**, *56*, 727–735. [CrossRef] [PubMed]
27. Zhdanova, E.Y.; Chubarova, N.Y.; Blumthaler, M. Biologically active UV-radiation and UV-resources in Moscow (1999–2013). *Geogr. Environ. Sustain.* **2014**, *7*, 71–85. [CrossRef]
28. Chubarova, N.E.; Pastukhova, A.S.; Galin, V.Y.; Smyshlyaev, S.P. Long-term variability of UV irradiance in the Moscow region according to measurement and modeling data. *Izv. Atmos. Ocean. Phys.* **2018**, *54*, 139–146. [CrossRef]
29. Čížková, K.; Láska, K.; Metelka, L.; Staněk, M. Reconstruction and analysis of erythemal UV radiation time series from Hradec Králové (Czech Republic) over the past 50 years. *Atmos. Chem. Phys.* **2018**, *18*, 1805–1818. [CrossRef]
30. Krzyścin, J.W.; Sobolewski, P.S. Trends in erythemal doses at the Polish Polar Station, Hornsund, Svalbard based on the homogenized measurements (1996–2016) and reconstructed data (1983–1995). *Atmos. Chem. Phys.* **2018**, *18*, 1–11. [CrossRef]
31. Román, R.; Bilbao, J.; de Miguel, A. Erythemal ultraviolet irradiation trends in the Iberian Peninsula from 1950 to 2011. *Atmos. Chem. Phys.* **2015**, *15*, 375–391. [CrossRef]
32. Litynska, Z.; Koepke, P.; De Backer, H.; Groebner, J.; Schmalwieser, A.; Vuilleumier, L. Long term changes and climatology of UV radiation over Europe. In *Final Scientific Report COST Action 726: UV Climatology for Europe*; European Cooperation in Science and Technology, COST 726 Project; 2010. Available online: http://www.meteoschweiz.admin.ch/web/en/research/completed_projects/cost_726.Par.0011.DownloadFile.tmp/finalreport.pdf (accessed on 1 May 2019).
33. Chubarova, N.Y. UV variability in Moscow according to long-term UV measurements and reconstruction model. *Atmos. Chem. Phys.* **2008**, *8*, 3025–3031. [CrossRef]
34. Fountoulakis, I.; Bais, A.F.; Fragkos, K.; Meleti, C.; Tourpali, K.; Zempila, M.M. Short- and long-term variability of spectral solar UV irradiance at Thessaloniki, Greece: Effects of changes in aerosols, total ozone and clouds. *Atmos. Chem. Phys.* **2016**, *16*, 2493–2505. [CrossRef]
35. Bais, A.F.; Tourpali, K.; Kazantzidis, A.; Akiyoshi, H.; Bekki, S.; Braesicke, P.; Chipperfield, M.P.; Dameris, M.; Eyring, V.; Garny, H.; et al. Projections of UV radiation changes in the 21st century: Impact of ozone recovery and cloud effects. *Atmos. Chem. Phys.* **2011**, *11*, 7533–7545. [CrossRef]
36. Watanabe, S.; Takemura, T.; Sudo, K.; Yokohata, T.; Kawase, H. Anthropogenic changes in the surface all-sky UV-B radiation through 1850–2005 simulated by an Earth system model. *Atmos. Chem. Phys.* **2012**, *12*, 5249–5257. [CrossRef]
37. Lamy, K.; Portafaix, T.; Josse, B.; Brogniez, C.; Godin-Beekmann, S.; Bencherif, H.; Revell, L.; Akiyoshi, H.; Bekki, S.; Hegglin, M.I.; et al. Clear-sky ultraviolet radiation modelling using output from the chemistry climate model initiative. *Atmos. Chem. Phys.* **2019**, *19*, 10087–10110. [CrossRef]
38. Egorova, T.; Rozanov, E.; Gröbner, J.; Hauser, M.; Schmutz, W. Montreal protocol benefits simulated with CCM SOCOL. *Atmos. Chem. Phys.* **2013**, *13*, 3811–3823. [CrossRef]
39. Pastukhova, A.S.; Chubarova, N.E.; Zhdanova, Y.Y.; Galin, V.Y.; Smyshlyaev, S.P. Numerical simulation of variations in ozone content, erythemal ultraviolet radiation, and ultraviolet resources over Northern Eurasia in the 21st century. *Izv. Atmos. Ocean. Phys.* **2019**, *55*, 242–250. [CrossRef]
40. McKenzie, R.L.; Liley, J.B.; Björn, L.O. UV radiation: Balancing risks and benefits. *Photochem. Photobiol.* **2009**, *85*, 88–98. [CrossRef]
41. Kinne, S. The Macv2 aerosol climatology. *Tellus B Chem. Phys. Meteorol.* **2019**, *71*, 1–21. [CrossRef]
42. Dee, D.P.; Uppala, S.M.; Simmons, A.J.; Berrisford, P.; Poli, P.; Kobayashi, S.; Andrae, U.; Balmaseda, M.A.; Balsamo, G.; Bauer, P.; et al. The ERA-Interim reanalysis: Configuration and performance of the data assimilation system. *Q. J. R. Meteorol. Soc.* **2011**, *137*, 553–597. [CrossRef]
43. Galin, V.Y.; Smyshlyaev, S.P.; Volodin, E.M. Combined chemistry-climate model of the atmosphere. *Izv. Atmos. Ocean. Phys.* **2007**, *43*, 399–412. [CrossRef]

44. Volodin, E.M.; Mortikov, E.V.; Kostrykin, S.V.; Galin, V.Y.; Lykossov, V.N.; Gritsun, A.S.; Diansky, N.A.; Gusev, A.V.; Iakovlev, N.G. Simulation of the present-day climate with the climate model INMCM5. *Clim. Dyn.* **2017**, *49*, 3715–3734. [CrossRef]
45. Chubarova, N.Y.; Nezval, Y.I. Ozone, aerosol and cloudiness impacts on biologically effective radiation and UV radiation less 380nm. In *Proceedings of the IRS'96 Current Problems in Atmospheric Radiation*; Smith, S., Ed.; A Deepak Publishing: Hampton, VA, USA, 1997; pp. 886–889.
46. Booth, C.R.; Madronich, S. Radiation amplification factors: Improved formulation accounts for large increases in ultraviolet radiation associated with antarctic ozone depletion. In *Ultraviolet Radiation in Antarctica: Measurements and Biological Effects*; American Geophysical Union (AGU): Washington, DC, USA, 2013; pp. 39–42. ISBN 978-1-118-66794-1.
47. Chubarova, N.; Zhdanova, Y.; Nezval, Y. A new parameterization of the UV irradiance altitude dependence for clear-sky conditions and its application in the on-line UV tool over Northern Eurasia. *Atmos. Chem. Phys.* **2016**, *16*, 11867–11881. [CrossRef]
48. Madronich, S.; Flocke, S. The role of solar radiation in atmospheric chemistry. In *Handbook of Environmental Chemistry Reactions and Processes*; Boule, P., Hutzinger, O., Eds.; Springer: Berlin/Heidelberg, Germany, 1998; Volume 2, pp. 1–26.
49. Rublev, A.; Trembach, V. 3D Monte-Carlo models and radiative transfer online tools. Intercomparison of three-dimensional radiation codes. In *Proceedings of the Three-Dimensional Radiation Codes: Abstracts of the First and Intercomparison Second International Workshops*; Cahalan, R.F., Davies, R., Eds.; University of Arizona Press: Tucson, AZ, USA, 2000; pp. 14–18. ISBN 0-9709609-0-5.
50. WMO (World Meteorological Organization). *Scientific Assessment of Ozone Depletion: 2006*; Global Ozone Research and Monitoring Project—Report No. 50; WMO: Geneva, Switzerland, 2007; p. 572.
51. Intergovernmental Panel on Climate Change (IPCC). Climate change 2001: The scientific basis. In *Contribution of Working Group I to the Third Assessment Report of the Intergovernmental Panel on Climate Change*; Houghton, J.T., Ding, Y., Griggs, D.J., Noguer, M., van der Linden, P.J., Dai, X., Maskell, K., Johnson, C.A., Eds.; Cambridge University Press: Cambridge, UK; New York, NY, USA, 2001; p. 881.
52. Intergovernmental Panel on Climate Change (IPCC). Climate change 2013: The physical science basis. In *Contribution of Working Group I to the Fifth Assessment Report of the Intergovernmental Panel on Climate Change*; Stocker, T.F., Qin, D., Plattner, G.-K., Tignor, M., Allen, S.K., Boschung, J., Nauels, A., Xia, Y., Bex, V., Midgley, P.M., Eds.; Cambridge University Press: Cambridge, UK; New York, NY, USA, 2013; p. 1535.
53. Dewolfe, W.A.; Wilson, A.; Lindholm, D.M.; Pankratz, C.K.; Snow, M.A.; Woods, T.N. Solar irradiance data products at the LASP interactive solar irradiance datacenter (LISIRD). *AGU Fall Meet. Abstr.* **2010**, *21*, GC21B-0881.
54. Thomason, L.; Peter, T. (Eds.) *SPARC: SPARC Assessment of Stratospheric Aerosol Properties (ASAP)*. SPARC Report No. 4, WCRP-124, WMO/TD—No. 1295. 2006. Available online: www.sparc-climate.org/publications/sparc-reports/ (accessed on 1 October 2019).
55. Rayner, N.A.; Parker, D.E.; Horton, E.B.; Folland, C.K.; Alexander, L.V.; Rowell, D.P.; Kent, E.C.; Kaplan, A. Global analyses of sea surface temperature, sea ice, and night marine air temperature since the late nineteenth century. *J. Geophys. Res. Atmos.* **2003**, *108*. Available online: https://agupubs.onlinelibrary.wiley.com/doi/10.1029/2002JD002670 (accessed on 1 October 2019). [CrossRef]
56. Stenke, A.; Schraner, M.; Rozanov, E.; Egorova, T.; Luo, B.; Peter, T. The SOCOL version 3.0 chemistry–climate model: Description, evaluation, and implications from an advanced transport algorithm. *Geosci. Model Dev.* **2013**, *6*, 1407–1427. [CrossRef]
57. Dragani, R. *On the Quality of the ERA-Interim Ozone Reanalyses. Part I: Comparisons with In Situ Measurements*; ERA Report Series, No. 2; ECMWF: Reading, UK, 2010. Available online: https://www.ecmwf.int/node/9111 (accessed on 11 October 2019).
58. Dragani, R. *On the Quality of the ERA-Interim Ozone Reanalyses. In Part II: Comparisons with Satellite Data*; ERA Report Series, No. 3; ECMWF: Reading, UK, 2010. Available online: https://www.ecmwf.int/node/9112 (accessed on 11 October 2019).
59. Dragani, R. On the quality of the ERA-Interim ozone reanalyses: Comparisons with satellite data. *Q. J. R. Meteorol. Soc.* **2011**, *137*, 1312–1326. [CrossRef]

60. Davis, S.M.; Hegglin, M.I.; Fujiwara, M.; Dragani, R.; Harada, Y.; Kobayashi, C.; Long, C.; Manney, G.L.; Nash, E.R.; Potter, G.L.; et al. Assessment of upper tropospheric and stratospheric water vapor and ozone in reanalyses as part of S-RIP. *Atmos. Chem. Phys.* **2017**, *17*, 12743–12778. [CrossRef]
61. Zhang, X.; Liang, S.; Wang, G.; Yao, Y.; Jiang, B.; Cheng, J. Evaluation of the reanalysis surface incident shortwave radiation products from NCEP, ECMWF, GSFC, and JMA using satellite and surface observations. *Remote Sens.* **2016**, *8*, 225. [CrossRef]
62. Träger-Chatterjee, C.; Müller, R.W.; Trentmann, J.; Bendix, J. Evaluation of ERA-40 and ERA-interim re-analysis incoming surface shortwave radiation datasets with mesoscale remote sensing data. *Meteorol. Z.* **2010**, *19*, 631–640. [CrossRef]
63. Szczypta, C.; Calvet, J.-C.; Albergel, C.; Balsamo, G.; Boussetta, S.; Carrer, D.; Lafont, S.; Meurey, C. Verification of the new ECMWF ERA-interim reanalysis over France. *Hydrol. Earth Syst. Sci.* **2011**, *15*, 647–666. [CrossRef]
64. *TOMS Science Team (Unrealeased), TOMS Nimbus-7 Total Ozone Aerosol Index UV-Reflectivity UV-B Erythemal Irradiances Daily L3 Global 1 deg × 1.25 deg V008*; Goddard Earth Sciences Data and Information Services Center (GES DISC): Greenbelt, MD, USA, 2017. Available online: https://disc.gsfc.nasa.gov/datacollection/TOMSN7L3_008.html (accessed on 17 January 2017).
65. *TOMS Science Team (Unrealeased), TOMS Earth-Probe Total Ozone (O3) Aerosol Index UV-Reflectivity UV-B Erythemal Irradiance Daily L3 Global 1 deg × 1.25 deg V008*; Goddard Earth Sciences Data and Information Services Center (GES DISC): Greenbelt, MD, USA, 2017. Available online: https://disc.gsfc.nasa.gov/datacollection/TOMSEPL3_008.html (accessed on 17 January 2017).
66. Jari, H.; Antii, A.; Johanna, T. *OMI/Aura Surface UVB Irradiance and Erythemal Dose Daily L3 Global Gridded 1.0 degree × 1.0 degree V3*; NASA Goddard Space Flight Center: Greenbelt, MD, USA; Goddard Earth Sciences Data and Information Services Center (GES DISC): Greenbelt, MD, USA, 2013. Available online: http://www.10.5067/Aura/OMI/DATA3009 (accessed on 16 January 2017).
67. Krotkov, N.A.; Herman, J.R.; Bhartia, P.K.; Seftor, C.J.; Arola, A.; Kaurola, J.; Kalliskota, S.; Taalas, P.; Geogdzhayev, I.V. Version 2 total ozone mapping spectrometer ultraviolet algorithm: Problems and enhancements. *Opt. Eng.* **2002**, *41*, 3028–3040.
68. Arola, A.; Kazadzis, S.; Krotkov, N.; Bais, A.; Gröbner, J.; Herman, J.R. Assessment of TOMS UV bias due to absorbing aerosols. *J. Geophys. Res. Atmos.* **2005**, *110*. Available online: https://agupubs.onlinelibrary.wiley.com/doi/10.1029/2005JD005913 (accessed on 15 November 2019). [CrossRef]
69. Bodeker, G.E.; Shiona, H.; Eskes, H. Indicators of Antarctic ozone depletion. *Atmos. Chem. Phys.* **2005**, *5*, 2603–2615. [CrossRef]
70. Zhdanova, Y.; Chubarova, N.; Nezval, Y. A method of estimating cloud transmission in the UV spectral range using data from different satellite measurements and reanalysis. In *AIP Conference Proceedings*; AIP: College Park, MD, USA, 2013; Volume 1531, pp. 911–914.
71. Fitzpatrick, T.B. The Validity and practicality of sun-reactive skin types I through VI. *Arch. Dermatol.* **1988**, *124*, 869–871. [CrossRef] [PubMed]
72. World Health Organization; World Meteorological Organization; United Nations Environment Programme, International Commission on Non-Ionizing Radiation Protection. *Global Solar UV Index: A Practical Guide: A JOINT RECOMMENDATION of World Health Organization*; WHO: Geneva, Switzerland, 2002; p. 28.
73. *Rationalizing Nomenclature for UV Doses and Effects on Humans; Joint Publication of CIE and WMO.* CIE 209:2014, WMO/GAW Report No. 211. 2014, p. 14. Available online: https://library.wmo.int/doc_num.php?explnum_id=7176 (accessed on 1 October 2019).
74. Holick, M.F.; Jenkins, M. *The UV Advantage: New Medical Breakthroughs Reveal Powerful Health Benefits from Sun Exposure and Tanning*; Ibook: New York, NY, USA, 2003.
75. Petropavlovskikh, I.; Godin-Beekmann, S.; Hubert, D.; Damadeo, R.; Hassler, B.; Sofieva, V.; Commission, I.O. *GAW Report, 241. SPARC/IO3C/GAW Report on Long-Term Ozone Trends and Uncertainties in the Stratosphere*; World Meteorological Organization (WMO): Geneva, Switzerland, 2018.
76. Petropavlovskikh, I.; Evans, R.; McConville, G.; Manney, G.L.; Rieder, H.E. The influence of the North Atlantic Oscillation and El Niño—Southern oscillation on mean and extreme values of column ozone over the United States. *Atmos. Chem. Phys.* **2015**, *15*, 1585–1598. [CrossRef]
77. Vigouroux, C.; Blumenstock, T.; Coffey, M.; Errera, Q.; García, O.; Jones, N.B.; Hannigan, J.W.; Hase, F.; Liley, B.; Mahieu, E.; et al. Trends of ozone total columns and vertical distribution from FTIR observations at eight NDACC stations around the globe. *Atmos. Chem. Phys.* **2015**, *15*, 2915–2933. [CrossRef]

78. Nezval, E.I.; Chubarova, N.E. Long-term variability of UV radiation in the spectral range of 300–380 nm in Moscow. *Russ. Meteorol. Hydrol.* **2017**, *42*, 693–699. [CrossRef]
79. Trenberth, K.E.; Fasullo, J.T. Global warming due to increasing absorbed solar radiation. *Geophys. Res. Lett.* **2009**, *36*. Available online: https://agupubs.onlinelibrary.wiley.com/doi/10.1029/2009GL037527 (accessed on 1 October 2019). [CrossRef]
80. Eyring, V.; Shepherd, T.G.; Waugh, D.W. (Eds.) *SPARC CCMVal Report on the Evaluation of Chemistry-Climate Models*. SPARC Report No. 5, WCRP-30, WMO/TD-No. 400. 2010. Available online: http://www.atmosp.physics.utoronto.ca/SPARC/ccmval_final/index.php (accessed on 1 October 2019).
81. Liu, Y.; Key, J.R.; Liu, Z.; Wang, X.; Vavrus, S.J. A cloudier Arctic expected with diminishing sea ice. *Geophys. Res. Lett.* **2012**, *39*. Available online: https://agupubs.onlinelibrary.wiley.com/doi/10.1029/2012GL051251 (accessed on 1 October 2019).

Article

Discontinuities in the Ozone Concentration Time Series from MERRA 2 Reanalysis

Peter Krizan *, Michal Kozubek and Jan Lastovicka

Institute of Atmospheric Physics, Czech Academy of Sciences, 14100 Prague, Czech Republic; kom@ufa.cas.cz (M.K.); jla@ufa.cas.cz (J.L.)

* Correspondence: krizan@ufa.cas.cz

Received: 17 October 2019; Accepted: 12 December 2019; Published: 14 December 2019

Abstract: Artificial discontinuities in time series are a great problem for trend analysis because they influence the values of the trend and its significance. The aim of this paper is to investigate their occurrence in the Modern-Era Retrospective analysis for Research and Applications, version 2 (MERRA 2) ozone concentration data. It is the first step toward the utilization of the MERRA 2 ozone data for trend analysis. We use the Pettitt homogeneity test to search for discontinuities in the ozone time series. We showed the data above 4 hPa are not suitable for trend analyses due to the unrealistic patterns in an average ozone concentration and due to the frequent occurrence of significant discontinuities. Below this layer in the stratosphere, their number is much smaller, and mostly, they are insignificant, and the patterns of the average ozone concentration are explainable. In the troposphere, the number of discontinuities increases, but they are insignificant. The transition from Solar Backscatter Ultraviolet Radiometer (SBUV) to Earth Observing System (EOS) Aura data in 2004 is visible only above 1 hPa, where the data are not suitable for trend analyses due to other reasons. We can conclude the MERRA 2 ozone concentration data can be used in trend analysis with caution only below 4 hPa.

Keywords: merra ozone data; discontinuities in reanalysis time series; trend analyses

---

1. Introduction

Ozone is a very important trace gas in the atmosphere because it protects the earth's biota from harmful ultraviolet radiation. The high scientific interest in atmospheric ozone arose near the beginning of the 1970s in connection with the problem of the possible supersonic transport impact on the ozone layer. After discovering the Antarctic ozone hole [1], the ozone became an important topic of atmospheric research. The origin of the ozone hole is in the chemical reactions of anthropogenic halogen radicals, which destroy ozone [2]. These results led to signing the Montreal Protocol and its amendments, which stopped the production of the ozone-depleting substances (ODS). This protocol has a positive influence on the ozone layer due to the decrease in ODS [3]. There are some signs of the ozone recovery, especially in the upper stratosphere [4]. In addition, all model studies predict future recovery of the ozone layer [5]. The concentration of ODS is not the only factor that influences the ozone layer. In addition, the greenhouse cooling of the stratosphere [6] and changes in the Brewer–Dobson circulation [7] have an impact on the ozone concentration. In such situations, proper trend analysis is necessary for the understanding of the ozone layer behaviour. Trend analysis was done from ground-based data [8,9]. These data are measured at one point, and the number of ground-based measurements of ozone is insufficient, especially in the Southern Hemisphere. Satellite ozone data have broader coverage, and these data are widely used in trend analysis in ozone research [10]. But in some areas, it is impossible to measure ozone (polar night, below dense clouds) by satellite.

On the other hand, the ozone data from reanalysis are temporally and spatially homogeneous, but there is a big question of the suitability of these data for trend analysis due to the occurrence of discontinuities in them [11]. They can be caused by satellite or instrument replacement or by assimilation of not homogenous basic parameters. [12] considered the evaluation of trends from reanalysis to be a challenge. The aim of this paper is to detect grid points in which discontinuities occur with the help of the Pettitt homogeneity test in the MERRA 2 ozone data in the period 1980–2017. Our study can help us to identify where it is possible to use the MERRA 2 reanalysis for trend study. This paper is divided into the following sections: Section 2 describes the data and method, Section 3 provides results, Section 4 discusses the results, and Section 5 concludes our results.

2. Data and Method

In this paper, we used MERRA 2 ozone monthly mean data from 500 hPa up to the top of the model (0.01 hPa) in the period 1980–2017. MERRA 2 reanalysis covers the whole satellite era. It includes substantial upgrades and changes to the data assimilation system and input data in comparison to previous MERRA reanalysis. New constraints are applied to ensure the conservation of global dry-air mass and to close the balance between surface water fluxes (precipitation minus evaporation) and changes in total atmospheric water [13]. The modified gravity wave scheme substantially improves the model representation of the quasi biennial oscillation (QBO) [14,15]. The assimilation of Microwave Limb Sounder (MLS) temperature retrievals at high-pressure levels (higher than 5 hPa) should improve the reanalysis at upper levels. The assimilation of MLS stratospheric ozone profiles and Ozone Monitoring Instrument (OMI) column ozone since the beginning of the Aura mission in late 2004 also improve the representation of fine-scale ozone features, especially in the region around the tropopause [16]. MERRA and MERRA 2 use the Three Dimensional Variational (3D VAR) assimilation process. MERRA 2 uses regular latitude–longitude grids from 1000 to 0.01 hPa (1/2° latitude × 5/8° longitude). Detail description of all reanalyses (included MERRA 2) can be found in ref [17,18].

We did our analysis for each grid point, we did not use any spatial (longitudinal and latitudinal) averages, because, during every averaging, some information is lost. In each grid and each layer, we used the time series of the ozone concentration and applied the Pettitt homogeneity test [19] to look for discontinuity in it. In each grid, the Pettitt test estimates only one main discontinuity, so this procedure is not able to detect multiple discontinuities. The Pettitt test is a nonparametric test. It has been developed to test homogeneity against the shift in a time series. In this approach, a shift point is at T; indicates that the time series can be divided into two subsequences x_t (t = 1..T) and x_t (t = T + 1,.. N). Thus, probability distribution functions $F_1(X)$ and $F_2(X)$ can be associated with the two subsequences. In practice, the null hypothesis H_0 is $F_1(X)$ is equal to $F_2(X)$, and the alternative hypothesis H_1 is $F_1(X)$ is not equal to $F_2(X)$. This method also involves a comparison of the observations so that:

$D_{i,j} = \mathrm{sgn}(x_i - x_j)$,

$D_{i,j} = 1$ if $x_i > x_j$,

$D_{i,j} = 0$ if $x_i = x_j$,

$D_{i,j} = -1$ if $x_i < x_j$.

We must compute the statistic:

$U_T, N = \sum_{t=1..T} \sum_{t=T,N} Di,j$, t = 1..N

Using the theory on statistic ranks, another K_N variable is derived from $U_{T,N}$ This new variable is defined:

K_N max abs($U_{T,N}$) T = 1..N − 1.

For the test, a probability of exceeding is fixed for a threshold value k given by the formula:

$P\ (K_{N>K}) \cong 6\exp(-6k^2)/(N^3 + N^2)$.

The null hypothesis, H_0 is rejected if the probability of exceedance given in P is less than the significance level α for a one-sided statistic test.

This test is widely used in the climatological research especially for precipitation and temperature analysis [20–22]. We searched for temporal occurrence of discontinuities, so in each grid with a discontinuity, we looked for the year of its occurrence, and we were able to detect the discontinuity occurrence in a certain year. We are interested in the spatial distribution of discontinuities as well. So in each layer and month, we constructed the map of their occurrence.

The Pettitt test tells us the year in which the discontinuity in time series occurs. But it does not say how big the discontinuity is or how it can affect trends. Small discontinuities have little impact on the trend analyses. On the other hand, the large one can have a strong trend impact, so we must divide the discontinuities according to their size. We tried to identify which ones were significant (in our case big enough to impact the trend) or insignificant according to this rule: Suppose we have a time series with the length L, and in year x, we observe the discontinuity. We compute the difference between the average before the year x and after this year. If this difference is larger than the variance of time series, we can say that the discontinuity is significant and will have some impact on trends, and we should be careful using this grid point in a trend analysis. This rule should be used mainly in areas with big variance because if the variance is very small, even the small discontinuities can be identified as significant. Another possibility for assessing the size of the discontinuities is the use of the Mann–Whitney test [23] for detecting significant ones. It would be very interesting to compare the results of both methods, and this topic could be solved in the future paper.

3. Results

3.1. Patterns of Ozone Concentration

In Figure 1, one can see the map of ozone concentration at 0.1 hPa (upper panel) and at 0.5 hPa (lower panel). In the uppermost model layer, we see the strange pattern of the ozone concentration, which cannot be considered as real. At 0.5 hPa, this strange pattern was also present, but we also see the maximum in ozone concentration above the Aleutian Islands. At 1 hPa in January (Figure 2 upper panel), the strange unrealistic pattern ceased, but the maximum in the ozone concentration appeared over the Aleutian Islands. In July (Figure 2 lower panel), this maximum was situated in the subpolar latitudes of the Southern Hemisphere. At 2 hPa, this maximum was weaker and at 3 hPa disappeared. In the upper stratosphere (Figure 3 upper panel), we observed the equatorial maximum and the polar minimum of ozone concentration. In this layer, the solar radiation is the main factor that drives the ozone concentration. In the lower stratosphere (Figure 3 lower panel), the maximal concentration of ozone was seen in the high latitudes and the minimal one over the equator, which is given by the Brewer–Dobson circulation. The lower stratospheric maximum was persistent in the upper troposphere (Figure 4 upper panel). At 500 hPa (Figure 4 lower panel), this tropospheric maximum disappeared, and we observed maximal ozone concentration over the polluted areas of China and the United States of America. So, we found several ozone concentration patterns from 500 hPa up to the top of the model: tropospheric pattern, lower stratospheric, middle stratospheric pattern, maximum pattern, and strange pattern in the uppermost model layers. The pressure ranges in which these patterns occurred are given in Table 1. The strange pattern occurred above 0.5 hPa. The pattern with a maximum above the subpolar latitudes was seen from 0.7 hPa down to 2 (3) hPa. From October to March, this maximum was situated in the Northern Hemisphere, while from April to October in the Southern Hemisphere. In October, both maxima were present, but the northern maximum was much stronger than the southern one. Below this pattern, the upper stratospheric one with the maximum over the equator and the minimum over high latitudes of both hemispheres could be seen. The lower limit of this pattern was about 20 hPa in each month, and the upper limit was monthly dependant (from 7 to 3 hPa). The upper limit for the lower stratospheric pattern was 30 (40) hPa in each month, while this pattern had its lower edge deep in the troposphere (Figure 4 upper panel) in the Northern Hemisphere and from December to June also in the Southern Hemisphere. From July to November, we saw a secondary ozone concentration maximum also in the subpolar latitudes of the Southern Hemisphere, which is given by the fact the ozone hole starts to form in polar latitudes and so this maximum appeared as a consequence of this formation. If the ozone-destroying reactions did not act, one could expect a higher concentration of ozone in the polar latitudes of the Southern Hemisphere, and so no subpolar maximum occurs.

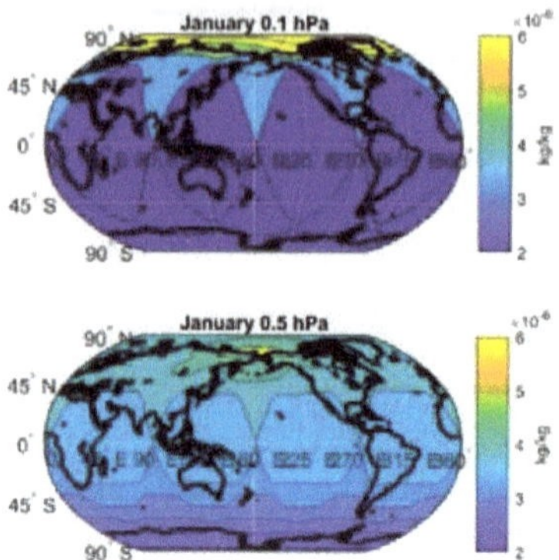

Figure 1. Average ozone concentration (kg/kg) in January at 0.1 hPa (**upper panel**) and at 0.5 hPa (**lower panel**).

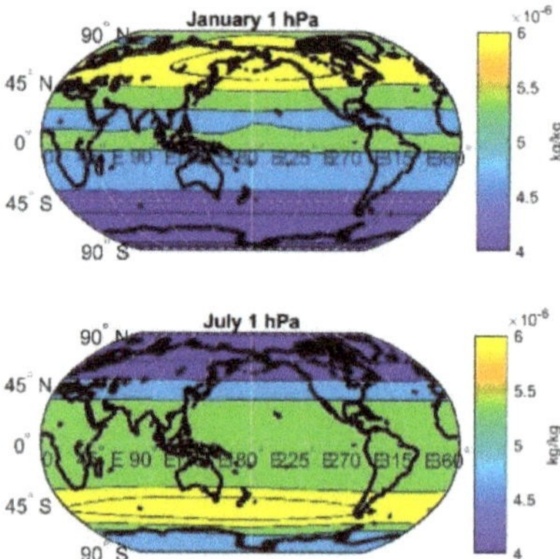

Figure 2. The same as Figure 1, but for January at 1 hPa (**upper panel**) and for July at 1 hPa (**lower panel**).

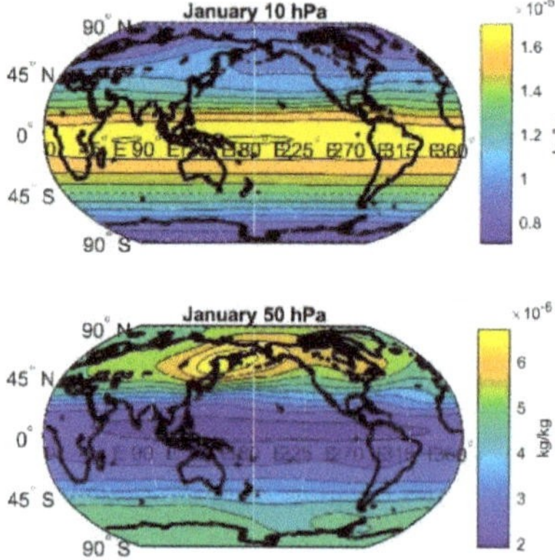

Figure 3. Average ozone concentration (kg/kg) in January at 10 hPa (**upper panel**) and at 50 hPa (**lower panel**).

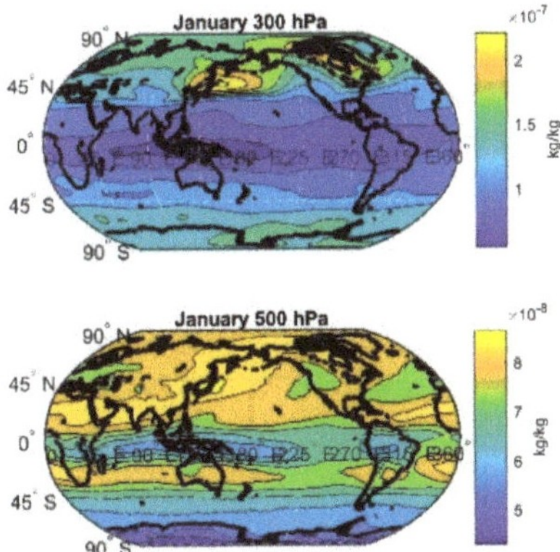

Figure 4. Average ozone concentration (kg/kg) in January at 300 hPa (**upper panel**) and at 500 hPa (**lower panel**).

Table 1. The vertical extent of ozone concentration pattern in a certain month (satellite—satellite-like pattern), max—the pattern with maximum over the Aleutian Islands (Al) or over Antarctica (An), UST—upper stratospheric pattern (maximum over equator and minimum over polar latitudes), LST—lower stratospheric pattern (maximum over the polar latitudes and minimum over equator), SH max—another maximum over the subpolar latitudes of the Southern Hemisphere.

	January	**February**	**March**	**April**	**May**	**June**
Satellite	0.1–0.5	0.1–0.5	0.1–0.5	0.1–0.5	0.1–0.5	0.1–0.5
max	0.7–2 / Al	0.7–2 / Al	2–3 / Al	0.7–2 / An	0.7–2 / An	0.7–3 / An
UST	3–10	3–10	5–10	7–20	7–20	7–20
LST	>30	>30	>30	>40	>40	>40
SH max	No	No	No	No	No	No
	July	**August**	**September**	**October**	**November**	**December**
Satellite	0.1–0.5	0.1–0.5	0.1–0.5	0.1–0.5	0.1–0.5	0.1–0.5
max	0.7–3 / An	0.7–3 / An	0.7–3 / An	0.7–2 / An,Al	0.7–2 / Al	0.7–2 /Al
UST	4–20	4–20	4–20	4–20	3–20	3–20
LST	>40	>40	>40	>40	>40	>40
SH max	30–250	30–250	30–250	30–300	30–300	No

3.2. *Temporal Occurrence of Discontinuities*

Now we deal with the temporal occurrence of discontinuities in the MERRA 2 ozone data. When we look at Table 2, we see in the uppermost model layer in each month the unimodal temporal distribution with a maximum in 1993 (U^{93} Figure 5—upper panel for January). Going down this unimodal distribution change, the bimodal one with the main maximum was observed in 1993, and the secondary maximum was observed in 2004($B^{\mathbf{93},04}$). In lower layers, the bimodal distribution with the main maximum in 2004 and the secondary one in 1993 ($B^{93,\mathbf{04}}$) could be seen. This transformation ended at 1 hPa (Figure 5 lower panel for January), where we observed unimodal distribution again, but the maximum occurred in 2004 in each month. Below 2 hPa down to 200 hPa, we observed flat temporal distribution of discontinuities (F) with no sharp maximum (Figure 6 upper panel for January) in each month. In January, the upper distribution border lay in 5 hPa. Below 200 hPa, we again observed unimodal temporal distribution with a maximum about the early 1990s (U^{93T}, Figure 6—lower panel for January), but this distribution was broader than distribution U^{93} in the upper model layers. When we observed the unimodal distribution of all discontinuities, the distribution of the significant ones must be similar. In the case of bimodal distribution, the situation was different. We could observe the bimodal distribution of all discontinuities, but the distribution of their significant counterparts can be unimodal. This was the case of bimodal distributions from 0.4 hPa and 0.5 hPa when we observed the bimodal distribution of all discontinuities, but the 1993 maximum was formed predominantly by the insignificant ones, and the 2004 maximum consisted predominantly of the significant discontinuities (Figure 7). Thus the distribution of the significant discontinuities in these cases was bimodal, but the maximum in 2004 was much stronger than that in 1993. Similarly, in the troposphere, the distribution U^{93T} was seen only for all discontinuities, but for the significant ones, we observed nearly flat distribution due to the fact the 1993 maximum consisted predominantly of the insignificant discontinuities (Figure 8).

Table 2. Temporal distribution of discontinuities at certain layers and certain months (U^{93}—unimodal distribution with maximum in 1993, $B^{93,04}$—bimodal distribution with two maxima in 1993 and 2004, the height of maxima was comparable, $B^{\mathbf{93},04}$—bimodal distribution, maximum in 1993 was higher than that in 2004, $B^{93,\mathbf{04}}$—bimodal distribution, maximum in 1993 was lower than that in 2004, U^{04}—unimodal distribution with maximum in 2004, F—flat distribution with no maxima, U^{93T}—unimodal distribution with wide maximum in 1993 which occurred in the troposphere).

[hPa]	January	February	March	April	May	June
0.1	U^{93}	U^{93}	U^{93}	U^{93}	U^{93}	U^{93}
0.3	$B^{\mathbf{93},04}$	$B^{\mathbf{93},04}$	$B^{\mathbf{93},04}$	$B^{\mathbf{93},04}$	U^{93}	U^{93}
0.4	$B^{93,04}$	$B^{93,04}$	$B^{93,\mathbf{04}}$	$B^{93,04}$	$B^{\mathbf{93},04}$	$B^{93,04}$
0.5	$B^{93,\mathbf{04}}$	$B^{93,\mathbf{04}}$	U^{04}	U^{04}	$B^{93,\mathbf{04}}$	$B^{93,\mathbf{04}}$
0.7	$B^{93,\mathbf{04}}$	U^{04}	U^{04}	U^{04}	U^{04}	U^{04}
1	U^{04}	U^{04}	U^{04}	U^{04}	U^{04}	U^{04}
2	U^{04}	U^{04}	U^{04}	U^{04}	F	F
3	U^{04}	F	F	F	F	F
4	U^{04}	F	F	F	F	F
5	U^{04}	F	F	F	F	F
7–200	F	F	F	F	F	F
250–500	U^{93T}	U^{93T}	U^{93T}	U^{93T}	U^{93T}	U^{93T}
	July	August	September	October	November	December
0.1	U^{93}	U^{93}	U^{93}	U^{93}	U^{93}	U^{93}
0.3	U^{93}	U^{93}	$B^{\mathbf{93},04}$	U^{93}	U^{93}	U^{93}
0.4	$B^{93,04}$	$B^{93,04}$	$B^{93,04}$	$B^{93,04}$	$B^{\mathbf{93},04}$	$B^{\mathbf{93},04}$
0.5	$B^{93,\mathbf{04}}$	$B^{93,\mathbf{04}}$	$B^{93,\mathbf{04}}$	$B^{93,\mathbf{04}}$	$B^{93,\mathbf{04}}$	$B^{93,04}$
0.7	$B^{93,\mathbf{04}}$	$B^{93,\mathbf{04}}$	U^{04}	U^{04}	U^{04}	$B^{93,\mathbf{04}}$
1	U^{04}	U^{04}	U^{04}	U^{04}	U^{04}	U^{04}
2	U^{04}	U^{04}	U^{04}	U^{04}	F	F
3	F	F	F	F	F	F
4	F	F	F	F	F	F
5	F	F	F	F	F	F
7–200	F	F	F	F	F	F
250–500	U^{93T}	U^{93T}	U^{93T}	U^{93T}	U^{93T}	U^{93T}

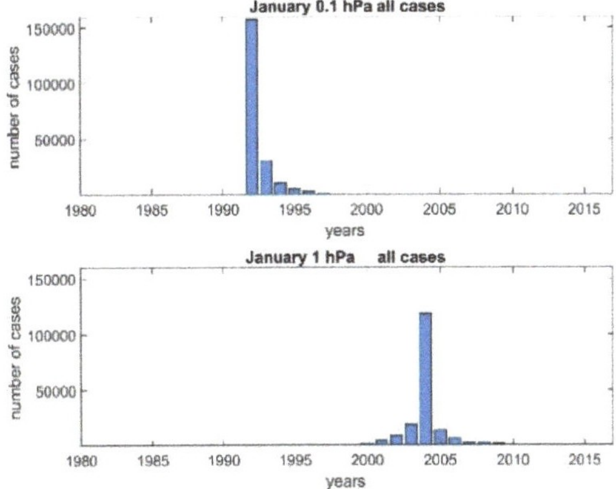

Figure 5. Temporal distribution of all discontinuities in January at 0.1 hPa (**upper panel**) and at 1 hPa (**lower panel**).

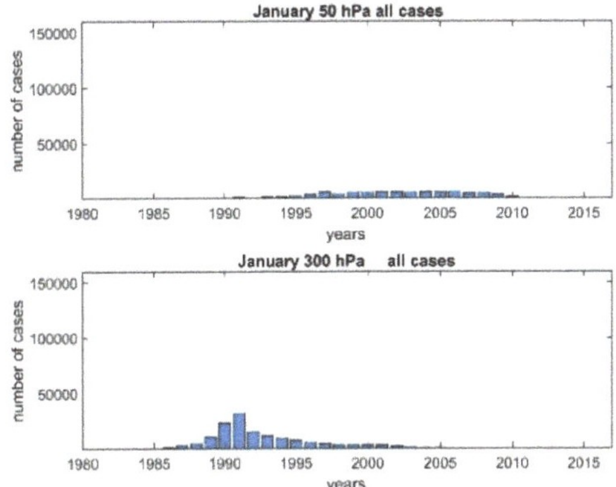

Figure 6. The same as Figure 5, but for 50 hPa (**upper panel**) and for 300 hPa (**lower panel**).

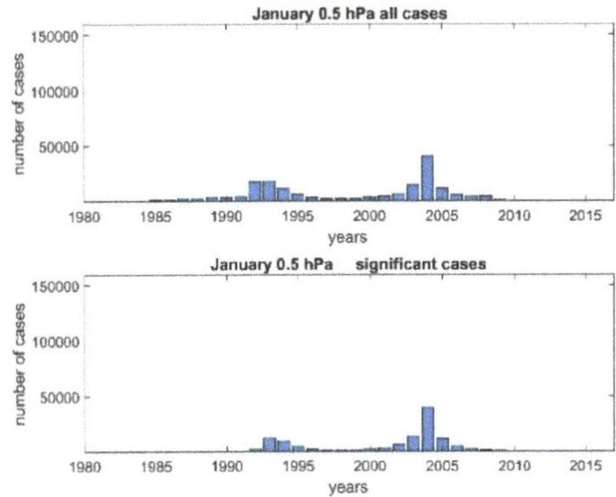

Figure 7. Temporal distribution at 0.5 hPa of all (**upper panel**) and significant (**lower panel**) discontinuities.

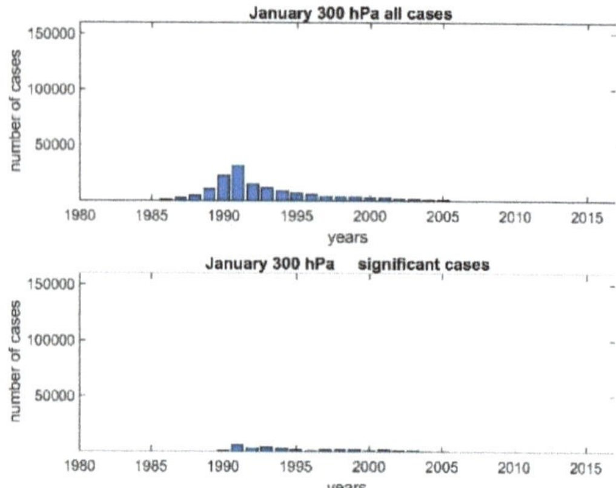

Figure 8. The same as Figure 7, but for 300 hPa.

3.3. *Vertical Profile of the Discontinuities Occurrence*

In Figure 9 (upper panel), we see the vertical profile of the percentage occurrence of all discontinuities. One hundred percent means that the discontinuity occurred in each grid point (207,936 points), and 0% means there were no discontinuities in time series over the globe. The vertical profile pattern was similar in each month. In the uppermost model layer at each grid, the discontinuity occurred. Going down, the share of all discontinuities (Figure 9 upper panel) declined to 0.7 hPa. In 1 hPa, we observed their local maximum. Below this layer, the number of discontinuities declined to about 10 hPa, where we observed local occurrence minimum. From 10 hPa down to 200 hPa, the discontinuities occurred in 40%–60% of all grids. In the troposphere, the share of all discontinuities increased.

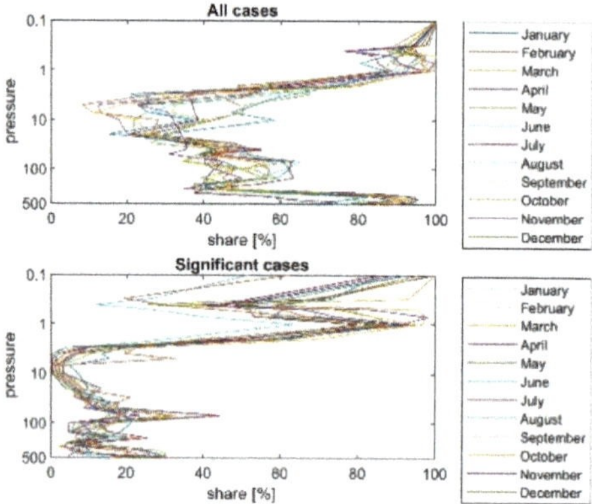

Figure 9. Vertical profiles of percentage occurrence of all (**upper panel**) and significant discontinuities (**lower panel**) in all months.

The vertical profiles of the significant discontinuities were similar (Figure 9 lower panel), but there were some interesting features in them. The local minimum about 0.7 hPa was much deeper than in the case of all discontinuities. The number of significant discontinuities was nearly zero about 10 hPa, and the tropospheric maximum, which was seen in the case of all discontinuities, was not present for the significant ones. With the increasing difference between the percentage of all discontinuities and the percentage of the significant ones, the share of significant discontinuities decreased.

3.4. Spatial Occurrence of Discontinuities

Now we deal with the spatial occurrence of discontinuities over the globe. We present the results for January, but the results for other months were similar. Figure 10 (upper panel) shows the geographical distribution of all discontinuities at 0.1 hPa, and in the lower panel, the distribution of the significant ones is displayed. In this layer, there were discontinuities in all grids (orange colour in the upper panel of Figure 10), and their reduction for the significant ones was small. At 3hPa (Figure 11), the situation was different. The number of all discontinuities (upper panel) was smaller, and the number of their significant counterparts (lower panel) was strongly reduced. From 5 hPa down to 250 hPa, the number of significant and insignificant discontinuities were relatively low, and at these layers, almost there were no significant discontinuities (Figures 12 and 13). From 300 hPa down to 500 hPa (Figure 14), the number of all discontinuities was very high, but the number of significant ones was low.

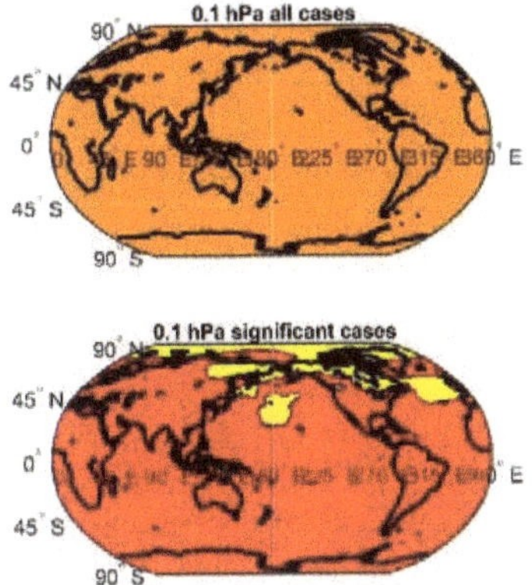

Figure 10. Geographical distribution of all (**upper panel**) and the significant (**lower panel**) discontinuities for January at 0.1 hPa (red—discontinuities, yellow—no discontinuities; upper panel all points have discontinuities.).

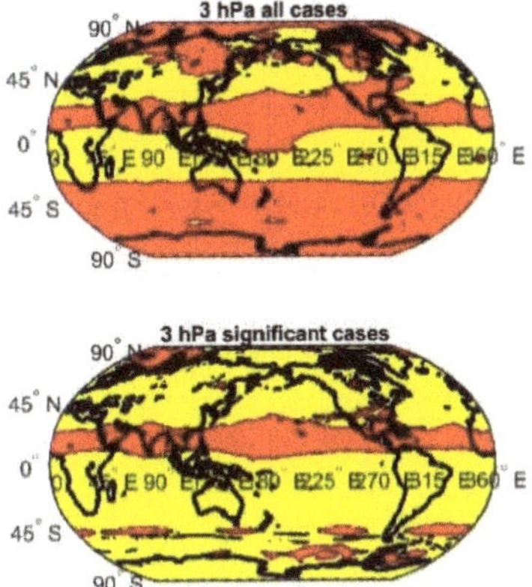

Figure 11. The same as Figure 10, but for 3 hPa.

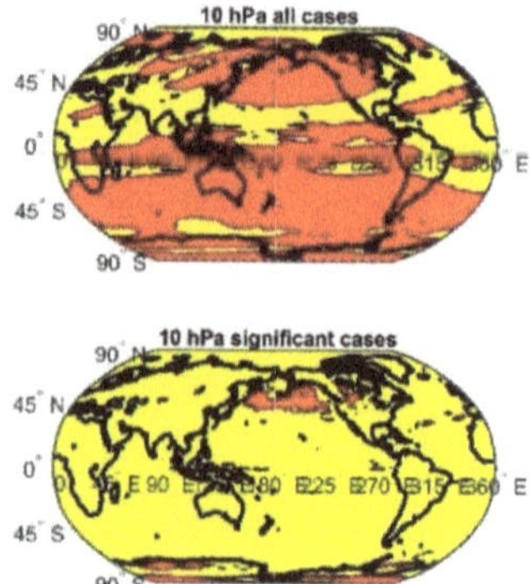

Figure 12. Geographical distribution of all (**upper panel**) and the significant (**lower panel**) discontinuities for January at 10 hPa.

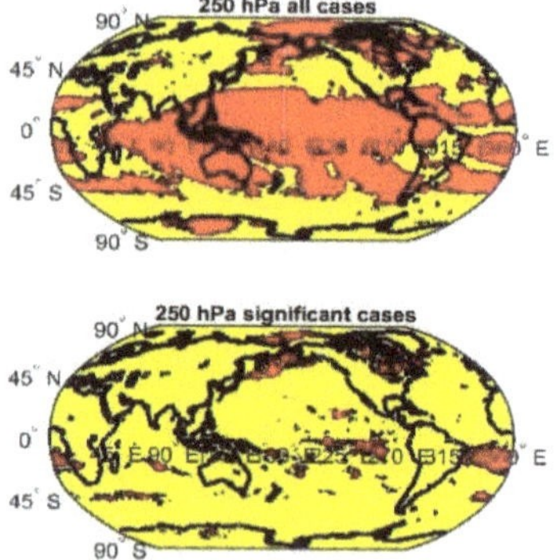

Figure 13. The same as Figure 12, but for 250 hPa.

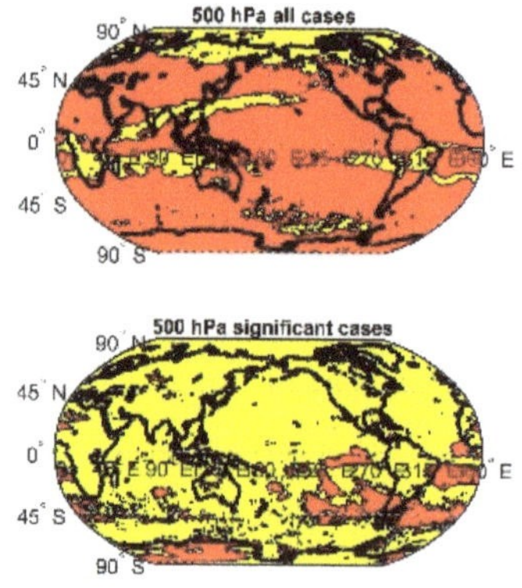

Figure 14. Geographical distribution of all (**upper panel**) and the significant (**lower panel**) discontinuities for January at 500 hPa.

Now we search geographical areas where the discontinuities occur more often. This analysis will be done separately for all and significant discontinuities. At each layer and each grid, we computed for all months the sum of cases when we observed discontinuity in a certain grid. When the sum equals 12, it means that in this grid, the discontinuity was observed in each month. On the other hand, when the sum equals 0, it means there was no discontinuity in this grid in every month. The higher the sum, the more frequent occurrence of discontinuities was seen in this grid. At the uppermost model layers, the spatial distribution of the sum was layer-dependant (from 0.1 hPa down to 1 hPa). This can be illustrated in Figures 15 and 16, where in the case of the insignificant discontinuities (lower panel) we observed at 0.3 hPa, a higher sum in the tropics than in the middle latitudes, but at 0.4 hPa (one layer below), the opposite was true. We observed a decrease in the sum from 1 hPa (Figure 17) down to 5 hPa (Figure 18) due to lower numbers of discontinuities. From 5 hPa down to 250 hPa, this sum for all and the significant discontinuities was relatively low, because the occurrence of discontinuities was rare. (Figure 19). In the troposphere (300 hPa Figure 20), we observed the increase in this sum, especially over the Pacific, in the case of all discontinuities. This sum was relatively low for the significant ones. This was caused by the vertical profile of discontinuity occurrence in the troposphere. From 300 hPa down to 500 hPa (Figure 21), the sum was high, but there were regions where this sum was much lower (over Africa and the Indian Ocean, over China, and over South America and the South Atlantic).

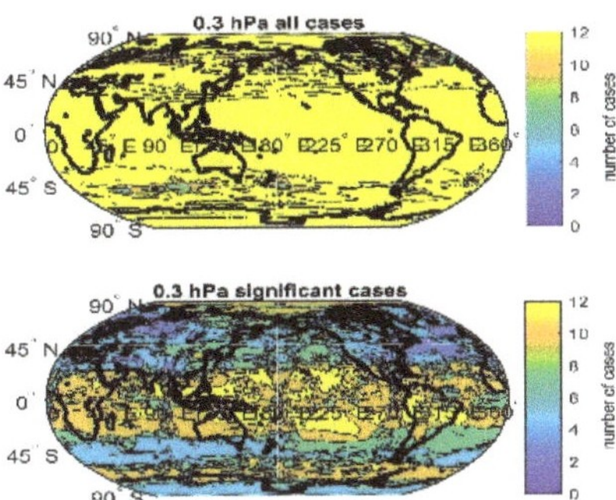

Figure 15. Geographical distribution of all (**upper panel**) and the significant (**lower panel**) discontinuities over the globe at 0.3 hPa.

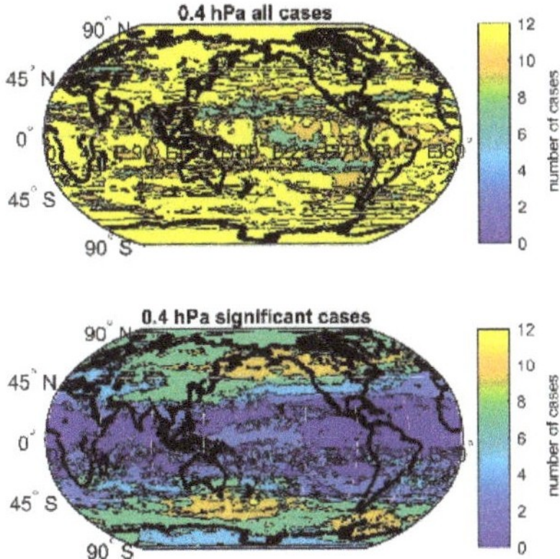

Figure 16. The same as Figure 15, but for 0.4 hPa.

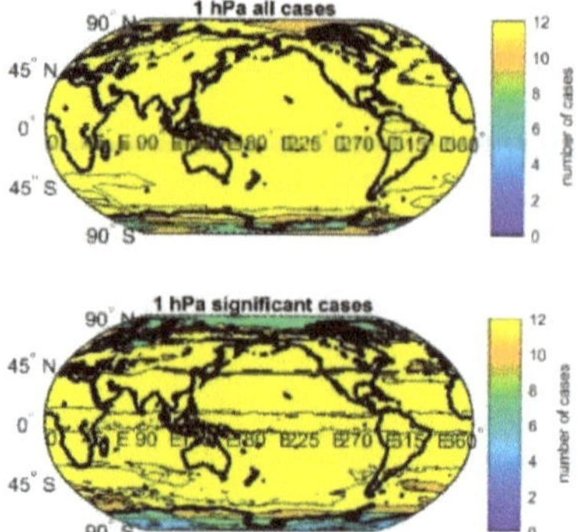

Figure 17. Geographical distribution of all (**upper panel**) and the significant (**lower panel**) discontinuities over the globe at 1 hPa.

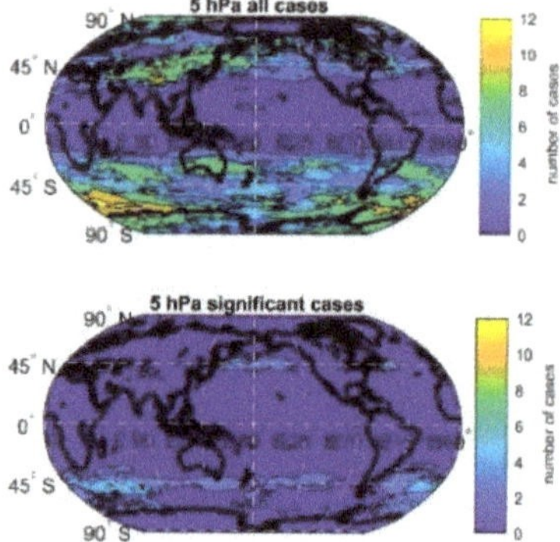

Figure 18. The same as Figure 17, but for 5 hPa.

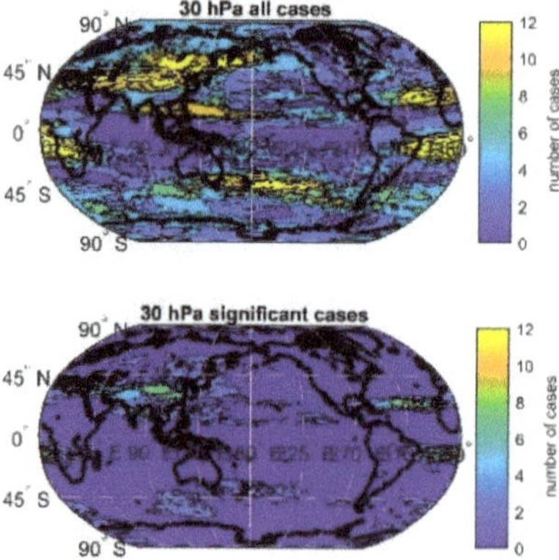

Figure 19. Geographical distribution of all (**upper panel**) and the significant (**lower panel**) discontinuities over the globe at 30 hPa.

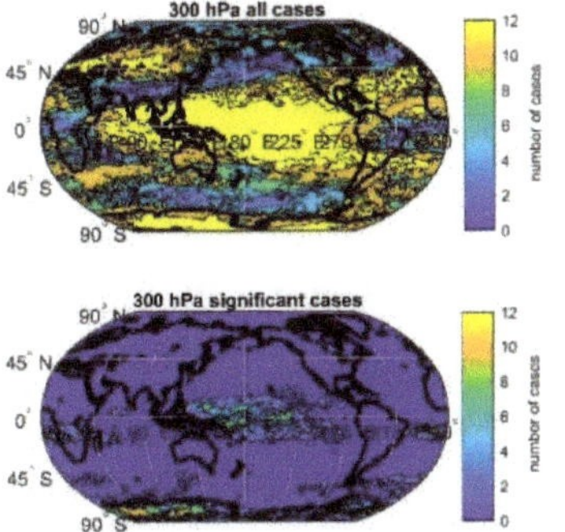

Figure 20. The same as Figure 19, but for 300 hPa.

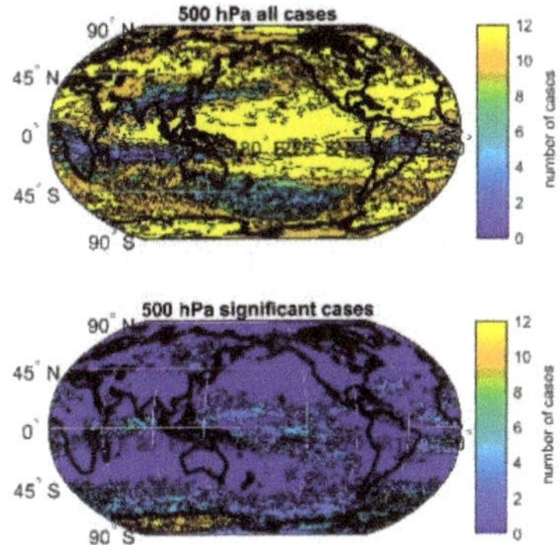

Figure 21. Geographical distribution of all (**upper panel**) and the significant (**lower panel**) discontinuities over the globe at 500 hPa.

4. Discussion

When we look at the ozone concentration patterns at the uppermost model layers, we see the strange patterns down to 0.5 hPa (Figures 1 and 2). According to our opinion, this is caused by the model itself, and these patterns are not real. So these layers are not suitable for trend analyses. The other groups of patterns are the ones with maximal concentration over the subpolar latitudes (from 0.7 to 2 hPa) of the winter hemisphere. It is very hard to explain this behaviour, because one can expect the main influence of solar activity to ozone at these heights, and in winter, there is very little solar radiation at the latitudes of maximum occurrence. In addition, at these heights, the ozone concentration is under the control of photochemistry, and the Brewer–Dobson circulation cannot act on them strongly. So one possible explanation is again the internal model variability, and the exploitation of these layers in trend analyses is problematic. Below these heights, the patterns with equatorial maximum and polar minimum were observed (from 3 down to 10 hPa). This is caused by the direct influence of solar radiation on ozone formation, which maximizes ozone in low latitude, and the Brewer–Dobson circulation is weak at these levels [24]. This upper stratospheric pattern can be explained by the theory, so in these heights, MERRA 2 gives reasonable results. From 30 hPa down to the tropopause, the maximal concentration was seen over high latitudes of both hemispheres and the minimal one over the equator. This is caused by the Brewer–Dobson circulation, which transports ozone rich air to high latitudes [25,26]. It is interesting these patterns persist deep into the troposphere, especially in the Northern Hemisphere, which may be caused by stratosphere–troposphere exchange in the vicinity of the Aleutian low.

The other topic of this paper, which is worth discussing, is the temporal occurrence of discontinuities in the period 1980–2017. In the uppermost model layer, the maximum of their occurrence was seen in 1993. One of the explanations of 1993 maximum is the Pinatubo eruption in 1991, but this maximum occurred only in the uppermost model layers. [27] found the enhanced wave activity in the two winters just after the Pinatubo eruption in similar heights. Thus the effect of the Pinatubo eruption may be real in the uppermost model layers. Going down this maximum disappeared, and the new one set up in 2004, and it was strongest at 1 hPa. This maximum can be clearly explained by the transition from SBUV to EOS Aura data in 2004 [12,28]. Below this layer, the flat temporal distribution of discontinuities was seen down to the troposphere, so no maximum in 2004 was observed. We can explain this fact by the following reasons. We can speculate the Pettitt test was either not suitable for this analysis, or the discontinuities in 2004 were not the most important in time series. The Pettitt test is widely used in atmospheric research, so we have no reason why this test does not work well for MERRA 2 ozone concentration data. If the 2004 discontinuities were not so huge, this fact could open the possibility of using MERRA 2 ozone data in trend analysis. The decision regarding these two hypotheses will be a matter of future research. In the troposphere, we observed wide maximum again in 1993, but not in 2004. In future work, we shall look at the wave activity in the upper troposphere and search its connection with the discontinuities occurrence peak in 1993.

Shangguan et al. [12] also detected discontinuities in 2015 when version 4.2 of the Microwave Limb Sounder (MLS) ozone data were used instead of version 2.2 and in 1998 when the Advanced TIROS (The Television Infrared Observation Satellite) Operational Vertical Sounder (ATOVS) was launched. In our paper, these discontinuities were not detected due to these reasons: We used data in the period of 1980–2017, so the year 2015 is at the edge of the series, and the Pettitt test could not capture this year. In each grid, we took into account only one main discontinuity, so the year 1998 could not be the main discontinuity, and thus, we did not notice it.

When we look at the vertical profile of discontinuity occurrence, we see the two main areas: the uppermost model layers and the tropospheric layers. From about 2 hPa down to the troposphere, we observed fewer discontinuities. High occurrence of discontinuities in the lowermost stratosphere and the upper troposphere can be explained by the lower quality of SBUV ozone data in these layers [29]. The share of the significant discontinuities was high in the upper model layer, while this share was low in the troposphere. The less number of (significant) discontinuities, the better for trend analysis.

We can conclude that due to ozone concentration patterns and the occurrence of discontinuities, the layers above 4 hPa are not suitable for trend analyses in MERRA 2. Below 4 hPa, the MERRA 2 gives explainable results expected from the theory, which also promotes using these data in the trend analysis. According to our results, the transition from SBUV to EOS Aura data does not have a major impact on discontinuity occurrence at these layers. The worst condition for trend analyses was the occurrence of the significant discontinuities in time series, and the share of grids with them was relatively low below 4 hPa. The better situation was the existence of discontinuities, which were not significant. Their number was higher than the significant ones. The grids with no discontinuities were not rare in the MERRA 2 ozone concentration data.

We must also deal with the possibility of the occurrence of discontinuities with natural origin. They must be included in trend analysis because they are real. It is very difficult to distinguish between artificial and real discontinuities because the real ones occur when there is some natural change in the atmosphere. We must consider the possibility of atmospheric changes connected to the change in ozone content and with increasing concentration of CO_2. Sudden water vapour drops in the stratosphere around 2000. In ref [30] can also be the manifestation of the changes in the atmospheric behaviour.

Almost all the modern reanalyses had a temporal discontinuity in 1998 when the ATOVS observations became available, and the reanalyses either switched immediately or transitioned from the TIROS Operational Vertical Sounder (TOVS) to the ATOVS over several years. These discontinuities were not consistent in each latitude or pressure level. The disagreement in the polar latitudes of the southern Hemisphere extended lower into the stratosphere than in the northern hemispheric ones.

The zonal wind variance was smaller than the temperature variance in the polar latitudes, but had a similar temporal difference between the TOVS and ATOVS time periods. For the tropics, the zonal wind variance is much larger than in the polar regions, and the disagreement between the semi-annual oscillation (SAO) and QBO zonal winds is quite large. Thus, improving equatorial winds in future reanalyses is an important goal. MERRA 2 reanalysis also showed discontinuities in 2004 when MLS observations were included. The disagreement was evident mainly for temperature or zonal wind. More details about discontinuities in reanalyses can be found in ref [31].

5. Conclusions

The main achievements concerning the ability to utilize the MERRA 2 ozone data for trend studies based on the Pettitt test results and on the analyses of monthly averages are the following:

- The data above 4 hPa are not suitable for the trend analyses due to occurrence of patterns of the average ozone concentration which cannot be explained by theory and due to the frequent occurrence of the significant discontinuities
- Below 4 hPa, the number of discontinuities is smaller, and they are mostly insignificant.
- In the troposphere, the number of discontinuities is higher than in the stratosphere, but they are mostly insignificant.
- The transition from SBUV to EOS Aura data in 2004 does not have a strong influence on discontinuities occurrence below 4 hPa.
- According to our results, the MERRA 2 ozone concentration data are suitable for trend analyses below 4 hPa with caution.

Author Contributions: This paper was done in close collaboration of the authors, but substantial part of the works was done by P.K.

Funding: Support by the Grant Agency of the Czech Republic via Grant 18-01625S is acknowledged.

Conflicts of Interest: The authors declare no conflict of interest.

References

1. Farman, J.; Gardiner, B.; Shanklin, J. Large losses of total ozone in Antarctica reveal seasonal ClOx/NOx interaction. *Nature* **1985**, *315*, 207–210. [CrossRef]
2. Solomon, S. Stratospheric ozone depletion: A review of concepts and history. *Rev. Geophys.* **1999**, *37*, 275–316. [CrossRef]
3. WMO. *Scientific Assessment of Ozone Depletion: 2014 Global Ozone Research and Monitoring Project Report*; World Meteorological Organization: Geneva, Switzerland, 2014; p. 416.
4. Harris, N.R.P.; Hassler, B.; Tummon, F.; Bodeker, G.E.; Hubert, D.; Petropavlovskikh, I.; Steinbrecht, W.; Anderson, J.; Bhartia, P.K.; Boone, C.D.; et al. Past changes in the vertical distribution of ozone—Part 3: Analysis and interpretation of trends. *Atmos. Chem. Phys.* **2015**, *15*, 9965–9982. [CrossRef]
5. Eyring, V.; Cionni, I.; Bodeker, G.E.; Charlton-Perez, A.J.; Kinnison, D.E.; Scinocca, J.F.; Waugh, D.W.; Akiyoshi, H.; Bekki, S.; Chipperfield, M.P.; et al. Multi-model assessment of stratospheric ozone return dates and ozone recovery in CCMVal-2 models. *Atmos. Chem. Phys.* **2010**, *10*, 9451–9472. [CrossRef]
6. Waugh, D.W.; Oman, L.; Kawa, S.R.; Stolarski, R.S.; Pawson, S.; Douglass, A.R.; Newman, P.A.; Nielsen, J.E. Impacts of climate change on stratospheric ozone recovery. *Geophys. Res. Lett.* **2009**, *36*, L03805. [CrossRef]
7. McLandress, C.H.; Sheepherd, T.G. Simulated anthropogenic changes in the brewer-dobson circulation, including its extension to high latitudes. *J. Clim.* **2009**, *22*, 1516–1540. [CrossRef]
8. Angell, J.K.; Melissa, F. Ground-based observations of the slowdown in ozone decline and onset of ozone increase. *J. Geophys. Res.* **2009**, *114*, D07303. [CrossRef]
9. Krzyscin, J.W.; Borkowski, J.L. Variabilty of the total ozone trend over Europe for the period 1950–2004 derived from reconstructed data. *Atmos. Chem. Phys.* **2008**, *8*, 2847–2857. [CrossRef]

10. Jones, A.; Urban, J.; Murtagh, D.P.; Eriksson, P.; Brohede, S.; Haley, C.; Degenstein, D.; Bourassa, A.; von Savigny, C.; Sonkaew, T.; et al. Evolution of stratospheric ozone and water vapour time series studied with satellite measurements. *Atmos. Chem. Phys.* **2009**, *9*, 6055–6075. [CrossRef]
11. Bengtsson, L.; Hagemann, S.; Hodges, K.I. Can climate trends be calculated from reanalysis data? *J. Geophys. Res. Atmos.* **2004**, *109*, D11111. [CrossRef]
12. Shangguan, M.; Wang, W.; Jin, S. Variability of temperature and ozone in the upper troposphere and lower stratosphere from multi-satellite observations and reanalysis data. *Atmos. Chem. Phys.* **2019**, *19*, 6659–6679. [CrossRef]
13. Takacs, L.L.; Suárez, M.J.; Todling, R. Maintaining atmospheric mass and water balance in reanalyses. *Q. J. Roy. Meteorol. Soc.* **2016**, *142*, 1565–1573. [CrossRef] [PubMed]
14. Molod, A.; Takacs, L.; Suarez, M.; Bacmeister, J. Development of the GEOS-5 atmospheric general circulation model: Evolution from MERRA to MERRA2. *Geosci. Model Dev.* **2015**, *8*, 1339–1356. [CrossRef]
15. Coy, L.; Wargan, K.; Molod, A.M.; McCarty, W.R.; Pawson, S. Structure and dynamics of the quasi-biennial oscillation in MERRA-2. *J. Clim.* **2016**, *29*, 5339–5354. [CrossRef] [PubMed]
16. Randles, C.A.; da Silva, A.M.; Buchard, V.; Darmenov, A.; Colarco, P.R.; Aquila, V.; Bian, H.; Nowottnick, E.P.; Pan, X.; Smirnov, A.; et al. *The MERRA-2 aerosol assimilation, NASA Technical Report Series on Global Modeling and Data Assimilation, Volume 45*; NASA/TM-2016-104606; NASA Goddard Space Flight Center: Greenbelt, MD, USA, 2016.
17. Fujiwara, M.; Wright, J.S.; Manney, G.L.; Gray, L.J.; Anstey, J.; Birner, T.; Davis, S.; Gerber, E.P.; Harvey, V.L.; Hegglin, M.I.; et al. Introduction to the SPARC Reanalysis Intercomparison Project (S-RIP) and overview of the reanalysis systems. *Atmos. Chem. Phys.* **2017**, *17*, 1417–1452. [CrossRef]
18. Gelaro, R.W.; McCarty, M.J.; Suárez, R.; Todling, A.; Molod, L.; Takacs, C.A.; Randles, A.; Darmenov, M.G.; Bosilovich, R.; Reichle, K.; et al. The modern-era retrospective analysis for research and applications, Version 2 (MERRA-2). *J. Clim.* **2017**, *30*, 5419–5454. [CrossRef]
19. Pettitt, A.N. A non-parametric approach to the change point detection. *Appl. Statist.* **1979**, *28*, 126–135. [CrossRef]
20. Javari, M. Trend and homogeneity analysis of precipitation in Iran. *Climate* **2016**, *4*, 44. [CrossRef]
21. Firat, M.; Dikbas, F.; Cem Koç, A.; Gungor, M. Missing data analysis and homogeneity test for Turkish precipitation series. *Sadhana* **2010**, *35*, 707–720. [CrossRef]
22. Wijngaard, J.B.; Klein Tank, A.M.G.; Knnen, G.P. Homogeneity of 20th century European daily temperature and precipitation series. *Int. J. Climatol.* **2003**, *23*, 679–692. [CrossRef]
23. Kruskal, W.H. Historical notes on the Wilcoxon unpaired two-sample test. *J. Am. Stat. Assoc.* **1957**, *52*, 356–360. [CrossRef]
24. Dütsch, H.U. The ozone distribution in the atmosphere. *Can. J. Chem.* **1974**, *52*, 1491–1504. [CrossRef]
25. Williams, R.S.; Hegglin, M.I.; Kerridge, B.J.; Jöckel, P.; Latter, B.G.; Plummer, D.A. Characterising the seasonal and geographical variability in tropospheric ozone, stratospheric influence and recent changes. *Atmos. Chem. Phys.* **2019**, *19*, 3589–3620. [CrossRef]
26. Rowland, F.S. Stratospheric ozone depletion. *Philos. Trans. R. Soc. Lond. B Biol. Sci.* **2006**, *361*, 769–790. [CrossRef] [PubMed]
27. Lastovicka, J.; Buresova, D.; Boska, J. Does the QBO and the Mt. Pinatubo eruption affect the gravity wave activity in the lower ionosphere? *Studia Geophys. Geod.* **1998**, *42*, 170–182. [CrossRef]
28. Wargan, K.; Labow, G.; Frith, S.; Pawson, S.; Livesey, N.; Partyka, G. Evaluation of the ozone fields in NASA's MERRA-2 reanalysis. *J. Clim.* **2017**, *30*, 2961–2988. [CrossRef] [PubMed]
29. Bhartia, P.K.; McPeters, R.D.; Flynn, L.E.; Taylor, S.; Kramarova, N.A.; Frith, S.; Fisher, B.; DeLand, M. Solar Backscatter UV (SBUV) total ozone and profile algorithm. *Atmos. Meas. Tech.* **2013**, *6*, 2533–2548. [CrossRef]
30. Brinkop, S.; Dameris, M.; Jöckel, P.; Garny, H.; Lossow, S.; Stiller, G. The millennium water vapour drop in chemistry-climate model simulations. *Atmos. Chem. Phys.* **2016**, *16*, 8125–8140. [CrossRef]
31. Long, C.S.; Fujiwara, M.; Davis, S.; Mitchell, D.M.; Wright, C.J. Climatology and interannual variability of dynamic variables in multiple reanalyses evaluated by the SPARC Reanalysis Intercomparison Project (S-RIP). *Atmos. Chem. Phys.* **2017**, *17*, 14593–14629. [CrossRef]

Article

Time Series Analysis and Forecasting Using a Novel Hybrid LSTM Data-Driven Model Based on Empirical Wavelet Transform Applied to Total Column of Ozone at Buenos Aires, Argentina (1966–2017)

Nkanyiso Mbatha [1,*] and Hassan Bencherif [2,3]

1 Department of Geography, University of Zululand, KwaDlangezwa 3886, South Africa
2 Laboratoire de l'Atmosphère et des Cyclones, LACy UMR 8105, Université de la Réunion, 97400 Réunion Island, France; hassan.bencherif@univ-reunion.fr
3 School of Chemistry and Physics, University of KwaZulu-Natal, Private Bag X54001, Durban 4000, South Africa
* Correspondence: nkanyisombatha5@gmail.com; Tel.: +27-035-902-6400

Received: 7 April 2020; Accepted: 25 April 2020; Published: 30 April 2020

Abstract: Total column of ozone (TCO) time series analysis and accurate forecasting is of great significance in monitoring the status of the Chapman Mechanism in the stratosphere, which prevents harmful UV radiation from reaching the Earth's surface. In this study, we performed a detailed time series analysis of the TCO data measured in Buenos Aires, Argentina. Moreover, hybrid data-driven forecasting models, based on long short-term memory networks (LSTM) recurrent neural networks (RNNs), are developed. We extracted the updated trend of the TCO time series by utilizing the singular spectrum analysis (SSA), empirical wavelet transform (EWT), empirical mode decomposition (EMD), and Mann-Kendall. In general, the TCO has been stable since the mid-1990s. The trend analysis shows that there is a recovery of ozone during the period from 2010 to 2017, apart from the decline of ozone observed during 2015, which is presumably associated with the Calbuco volcanic event. The EWT trend method seems to have effective power for trend identification, compared with others. In this study, we developed a robust data-driven hybrid time series-forecasting model (named EWT-LSTM) for the TCO time series forecasting. Our model has the advantage of utilizing the EWT technique in the decomposition stage of the LSTM process. We compared our model with (1) an LSTM model that uses EMD, namely EMD-LSTM; (2) an LSTM model that uses wavelet denoising (WD) (WD-LSTM); (3) a wavelet denoising EWT-LSTM (WD-EWT-LSTM); and (4) a wavelet denoising noise-reducing sequence called EMD-LSTM (WD-EMD-LSTM). The model that uses the EWT decomposition process (EWT-LSTM) outperformed the other five models developed here in terms of various forecasting performance evaluation criteria, such as the root mean square error (RMSE), mean absolute error (MAE), mean absolute percentage error (MAPE), and correlation coefficient (R).

Keywords: total column of ozone (TCO); trend estimates; long short-term memory networks (LSTM); empirical wavelet transform (EWT); forecasting; Mann-Kendall

1. Introduction

Total column of ozone (TCO), defined as the vertical integration of ozone number density profile in the lower and middle atmosphere, is important for regulating the amount of harmful ultraviolet (UV) radiation reaching the surface of the Earth. The decline of global stratospheric ozone column observed throughout the 1980s [1] and the discovery of the Southern Hemisphere polar region ozone hole [2,3] raised the awareness of the need to protect the global ozone layer. The ozone decline and Antarctic ozone hole, which is stronger during the Southern Hemisphere spring, is known to be the result of chemical reactions involving chlorine and bromine, which causes ozone in the southern polar region to be severely depleted. However, the Montreal Protocol, a global agreement initiated in 1987, became a binding agreement designed to protect the ozone layer by phasing out the production of ozone-depleting gases such as chlorofluorocarbons (CFCs).

A number of atmospheric chemistry research groups have been closely monitoring the slow self-recovery of stratospheric ozone for some time. The difficulty of this slow recovery is that the entire process depends solely on the self-recovery aspects of the ozone layer. Consequently, the recovery of the stratospheric ozone layer is strongly dependent on the continued decline in the atmospheric concentration of ozone-depleting gases such as CFC [4]. However, it is concerning that a recent study by Rigby et al. [4] reported that the recent slowing down of the stratospheric ozone layer recovery is counterbalanced by the continual emission of trichlorofluoromethane (CFC-11) in northeast China. The suggestion of a decrease in stratospheric ozone recovery and the continuation of the ozone decline in the lower stratosphere has been presented by other authors [5]. The study by Ball et al. [5] indicated that, while the stratospheric ozone layer has stopped declining across the globe, there is no clear increase observed at latitudes between 60° S and 60° N outside the polar region (60°–90°). Therefore, it is for this reason that TCO time series studies and the consequent design of models that can predict and forecast the dynamics of the ozone concentration in the atmosphere are imperative.

Studies of the time series analysis of satellite and ground-based TCO data revealed a significant decline of −3% to −6% per decade (depending on latitude) throughout the 1980s and 1990s. These were linked to CFCs increases in the atmosphere [6,7]. Chehade at al. [8] investigated the long-term evolution of zonal mean annual mean total ozone measurements (between 65° S and 65° N) from merged data sets of various satellites over a period of 1979–2012. Among other things, their study showed (1) a strong effect of the El Chichón (1982) and Mt. Pinatubo (1991) volcanic eruptions in the global TCO distribution and (2) evidence for a large fraction of ozone loss, which was more prominent in the Northern Hemisphere. Recently, Ball et al. [9] assessed the trend in stratospheric ozone using space-based global ozone observations from 1985 to 2018. Their study demonstrated that large inter-annual mid-latitude variations observed during 2017 resurgence are driven by non-linear quasi-biennial oscillation (QBO) phase-dependent seasonal variability. A year later, Ball et al. [5] quantified the absolute changes in ozone within different regions of the stratosphere and troposphere and speculated, by using a robust regression analysis approach and dynamical linear modeling, on the latitudinal contributions to the total column ozone since 1998. Utilizing multiple satellite data sets in their study, they managed to show evidence that ozone in the lower stratosphere between 60° S and 60° N has continued to decline since 1998.

Data-driven time series forecasting is an important research topic in the domain of science and engineering. The primary objective of this data science domain is to use available data to develop a mathematical model that can forecast future situations. Over the years, there have been many efforts in the development and improvement of time series forecasting models. These models can be classified into statistical models, artificial intelligence models, physical models, and hybrid models [10]. Statistical models like autoregressive integrated moving average (ARIMA) linear models [11] and artificial intelligence models like artificial neural networks (ANN) non-linear models [11,12] are widely popular data-driven time series models. However, a hybrid model [13], which is built by combining the two methods (non-linear and linear), has been shown to produce better results than the individual models [11,14–16]. To improve the performance of the hybrid models, signal decomposition is an important step when contracting the model [17–20]. The most popular methods for signal decomposition include discrete wavelet transform (DWT) [16,19], empirical mode decomposition (EMD), and ensemble empirical mode decomposition (EEMD) [11,21], among others. Recently, a new novel signal decomposition technique was developed, named the empirical wavelet transform (EWT) [22]. This technique integrates systems with a mathematical theory like wavelet transform (WT) and adaptiveness to the signal like EMD. Subsequently, this technique has proven to be efficient in the improvement of the performance of hybrid time series forecasting models [17,18,20].

Long short-term memory networks (LSTM) recurrent neural networks (RNNs), which were proposed by Hochreiter and Schmidhuber [23], have been proven as an enhanced variant of RNNs that can learn the information contained in time series data more robustly. LSTM, a deep learning method, can more effectively capture the variability of time series data compared with other models such as ARIMA and ANN. However, the model accuracy is not high, using the LSTM network in its simplest form, to predict the time series data. Therefore, applying it in its simplest form, in a non-stationary time series data like the TCO time series used here, consideration should be given to first decomposing the time series to several more predictable components, each having less non-stationarity in order to improve overall accuracy. LSTM has been previously used in digital currency forecasting [18], daily land surface temperature forecasting [11], wind speed forecasting [20], and wind power short-term prediction [19], among others.

In this paper, the analysis of the growth rate and trend of the long-term data series of the TCO measured at Buenos Aires for the period from 1966 to 2017 is carried out. This data is one of the longest recorded time series of TCO in the Southern Hemisphere. A number of studies have indicated that the EMD technique can be an important method for trend identification [10–12]. This research proposes the use of the EWT technique for the first time to identify the trend of the TCO time series. In addition, the authors aim to develop a robust data-driven hybrid model that uses the EWT technique in the decomposition stage of the LSTM, namely EWT-LSTM, and use this model in forecasting the TCO time series. The EWT-LSTM is subsequently compared to an LSTM model that uses EMD, namely EMD-LSTM. The study also develops the LSTM model that uses wavelet denoised (WD) data (WD-LSTM), wavelet denoising noise reduction EWT-LSTM (WD-EWT-LSTM), and wavelet denoising EMD-LSTM (WD-EMD-LSTM). Finally, statistical evaluation metrics (i.e., root mean square error (RMSE), mean absolute error (MAE), mean absolute percentage error (MAPE), and R) were used to assess the performance of the above-mentioned models, including a single LSTM model.

2. Materials and Methods

2.1. Total Column Ozone

The total column of ozone (TCO) data used in this paper are the long record of Dobson observations, measured at Observatorio Central Buenos Aires (OCBA) (34.60° S, 58.38° W) for the period from 1966 to 2017 [24], which was downloaded from the World Ozone and Ultraviolet Radiation Data Centre website (https://woudc.org/home.php). The Dobson spectrometer instrument was developed by G. M. B. Dobson [25]. Since then, this instrument played an instrumental role in the measurements of TCO in different locations around the Earth. It is also important to mention here that the first observation of the Antarctic ozone hole was observed by a Dobson instrument [2], and later confirmed by total ozone mapping spectrometer (TOMS) instrument [26]. The Dobson measurement method is based on the use of ozone absorption in the Huggins band. Owing to the possible interference by aerosols and molecules, the differential absorption at two wavelength pairs is measured to separate the ozone absorption from aerosols and molecules in the atmosphere. More details about Dobson spectrometers and its accuracy assessment can be found in a study by Grant [27].

2.2. Mann-Kendal

It is always vital to work out monotonic trends in the time series of any geophysical data before any advance use. In this study, the Mann-Kendall test [28,29] was used to detect the trends that exist among the used variables. This method is defined as a non-parametric, rank-based method that is commonly used to extract monotonic trends in the time series of climate data, environmental data, or hydrological data. The following formula (Equation (1)) is used to compute Mann-Kendall test statistics:

$$S = \sum_{k=1}^{n-1} \sum_{j=k+1}^{n} sgn\left(X_j - X_k\right) \tag{1}$$

where

$$sgn(x) = \begin{cases} +1, & \text{if } x > 1 \\ 0, & \text{if } x = 0 \\ -1, & \text{if } x < 1 \end{cases} \quad (2)$$

The average value of S is $E[S] = 0$, and the variance σ^2 is calculated using the following equation (Equation (3)):

$$\sigma^2 = \left\{ n(n-1)(2n+5) - \sum_{j=1}^{p} t_j\left(t_j - 1\right)\left(2t_j + 5\right) \right\} / 18 \quad (3)$$

where t_j is the number of data points in the jth tied group, and p is the number of the tied group in the time series. Under the assumption of a random and independent time series, the statistical function S is approximately normal distributed given that the below Z-transformation equation (Equation (4)) is used:

$$Z = \begin{cases} \frac{S-1}{\sigma} & \text{if } S > 1 \\ 0 & \text{if } S = 0 \\ \frac{S+1}{\sigma} & \text{if } S < 1 \end{cases} \quad (4)$$

The value of the statistic S is associated with Kendall's $\tau = \frac{S}{D}$, where

$$D = \left[\frac{1}{2}n(n-1) - \frac{1}{2}\sum_{j=1}^{p} t_j\left(t_j - 1\right) \right]^{1/2} \left[\frac{1}{2}n(n-1) \right]^{1/2} \quad (5)$$

In respect to the above-defined Z-transformation equation, this study reflects a 95% confidence level, where the null hypothesis of no trend is rejected if $|Z| > 1.96$. Another important output of the Mann-Kendall statistic is the Kendall tau "τ", which is a measure of a correlation, which measures the strength of the relationship between any two independent variables. In this study, the Mann-Kendall test system summarized above was applied to the TCO time series under scrutiny.

The limitations of the Mann-Kendall test are that it does not give the complete structure of the trend for the whole time series. Understanding the dynamics of the trend throughout the time series is important as there might be fluctuations in the trend over the investigation period. This limitation is solved by applying the test sequentially for every individual period. This specific version of the Mann-Kendall test statistic is called sequential Mann-Kendall (SQ-MK) [30]. The SQ-MK test method generates two time series, a forward/progressive one (u(t)) and a backward/retrograde one (u'(t)), and is furthermore capable of detecting approximate potential trend turning points in long-term time series, which can be obtained by plotting the progressive and retrograde time series in the same figure. If they happen to cross each other and diverge beyond the specific threshold (±1.96 in this study), then there is a statistically significant trend. To compute SQ-MK, rank values are used. The ranked values of y_i of a given time series $(x_1, x_2, x_3, \ldots, x_n)$ in the analysis, as well as the magnitudes of y_i, $(i = 1, 2, 3, \ldots, \text{n})$, are compared with y_i, $(j = 1, 2, 3, \ldots, \text{j} - 1)$. At each comparison, the number of cases where $y_i > y_j$ is counted and then donated to n_i. The statistic t_i is thus calculated by the following:

$$t_i = \sum_{j=1}^{i} n_i \quad (6)$$

The mean and variance of the statistic t_i are given by the following:

$$E(t_i) = \frac{i(i-1)}{4} \quad (7)$$

and

$$\text{Var}(t_i) = \frac{i(i-1)(2i-5)}{72} \tag{8}$$

The forward/progressive sequential values of the statistic $u(t_i)$ that are standardized are calculated using the following Equation (30):

$$u(t_i) = \frac{t_i - E(t_i)}{\sqrt{\text{Var}(t_i)}} \tag{9}$$

In order to calculate the backward/retrograde statistic values $u'(t_i)$, the same time series $(x_1, x_2, x_3, \ldots, x_n)$ is used, but statistic values are computed by starting from the end of the time series. The plotting of the forward and backward sequential statistic in the same figure allows for the detection of the approximate beginning of a developing trend. It is noteworthy to also mention that, in the SQ-MK trend analysis, sometimes the forward and backward trends cancel each other and finally do not give a significant trend. This method has been successfully employed by many researchers to detect trend turning points and its significance [28,30]. The Mann-Kendall trend test and the SQ-MK trend analysis were performed using package 'trend' [31] and 'pheno' [32], respectively, run in R version 3.5.0.

2.3. Empirical Mode Decomposition (EMD)

The empirical mode decomposition is a fundamental part of the Hilbert–Huang transform (HHT). The signal is decomposed into so-called intrinsic mode functions (IMFs) that include both the trend and finite oscillations that supply information on different scales of the original signal. This method is a self-adaptive time–space analysis method, designed for processing time series that are non-stationary and non-linear. In mathematical operation, the EMD decomposes the time series as a finite sum of $N+1$ IMFs $f_k(t)$, such that,

$$f(t) = \sum_{k=0}^{N} f_k(t). \tag{10}$$

An IMF is an amplitude modulated-frequency modulated function, which can be mathematically represented in the following form:

$$f_k(t) = F_k(t)\cos(\varphi_k(t)) \text{ where } F_k(t),\ \varphi'_k(t) > 0\ \ \forall t. \tag{11}$$

Following a study by Gilles [22] where they explored the EMD set of equations, in the above equation, it is assumed that f_k and φ'_k vary much slower than φ_k. It should also be noted that the IMF parameter f_k behaves as a harmonic component. These IMFs are extracted by first computing the upper, $\overline{f}(t)$, and the lower, $\underline{f}(t)$, envelopes using a cubic spline interpolation method from the maxima and minima of f. Thereafter, the mean envelop is calculated using the following formula:

$$m(t) = \frac{\left(\overline{f}(t) + \underline{f}(t)\right)}{2}, \tag{12}$$

and, finally, the IMF candidate is computed as follows:

$$r_1(t) = f(t) - m(t). \tag{13}$$

In normal circumstances, $r_1(t)$ does not fulfill the properties of an intrinsic mode function; hence, an acceptable IMF candidate is reached by iterating the same process to r_1 and the subsequent r_k. The ultimate retained IMF is given by the following:

$$f_1(f) = r_n(f), \tag{14}$$

Moreover, the next IMF is obtained by the same algorithm applied on $f(t) - f_1(t)$. The remaining IMFs are also extracted by repeating this algorithm on the successive residuals. While this algorithm seems to be highly adaptable and able to compute the non-stationary part of the original function, its main problem is that it is based on an ad-hoc process that is mathematically difficult to model [22]. This method also presents limitations when some noise is added in the signal. To solve some of these limitations, an ensemble EMD (EEMD) was proposed in a study by Torres et al. [33]. The key idea of the EEMD relies on averaging the modes obtained by EMD applied to several realizations of Gaussian white noise added to the original signal. The resulting decomposition solves the EMD mode-mixing problem; however, it introduces new ones. In the method proposed here, a particular noise is added at each stage of the decomposition and a unique residue is computed to obtain each mode. The resulting decomposition is complete, with a numerically negligible error. During a study by Gilles [22], this method was tested using two examples, namely a discrete Dirac delta function and an electrocardiogram signal. The results showed the following: (1) compared with EEMD, the new method (EWT) provides a better spectral separation of the modes; and (2) it requires a smaller number of sifting iterations, which reduces the computational cost. It should be noted that a study by Torres et al. [33] showed that the EEMD method does seem to stabilize the obtained EMD decomposition, but with an increase in computational cost. This is one of the reasons this study explored the used of the more theoretical driven EWT method, which is explained in a study by Gilles [22].

2.4. Empirical Wavelet Tranform

Our study proposes the use of a newely developed method (EWT) that is used to decompose any signal into a collection of amplitude modulated-frequency modulated (AM–FM) signals according to the information contained within the Fourier spectrum of the given signal. The method was developed by Gilles [22], and it enables the extraction of different modes of the signal by designing a suitable filter bank. This method is called the empirical wavelet transform. While the EMD is a traditional method that is normally used to decompose a signal according to its contained information within a signal, the main concern about this method is that it lacks mathematical theory. However, the EWT with its well-defined mathematical theory has been shown to perform better than the EMD (e.g., Gilles [22]). When applying the EWT in a time series $x(t)$, the process of signal decomposition can be described in the following five steps:

The first step is to calculate the Fourier spectrum $F(\omega)$ of the given signal using the fast Fourier transform algorithm.

The second step includes determining the boundaries ω_i by proper segmentation of the Fourier spectrum:

$$\omega_i = \frac{f_i + f_{i+1}}{2} \text{for } 1 \leq i \leq 1 \tag{15}$$

where $\{f_i\}$, $i = 1, 2, \ldots, N$ represents the frequencies corresponding to local maxima and $f_0 = 0$. The third step is to construct the empirical wavelet $\psi_i(\omega)$ and the empirical scaling function $\varphi_i(w)$ in the following form:

$$\varphi_i(\omega) = \begin{cases} 1, \text{ if } (1+\gamma)\omega_i \leq |\omega| \leq (1-\gamma)\omega_{i+1} \\ \cos\left[\frac{\pi}{2}\beta\left(\frac{1}{2\gamma\omega_{i+1}}(|\omega| - (1-\gamma)\omega_{i+1})\right)\right], \text{ if } (1-\gamma)\omega_{i+1} \leq |\omega| \leq (1+\gamma)\omega_{i+1} \\ \sin\left[\frac{\pi}{2}\beta\left(\frac{1}{2\gamma\omega_i}(|\omega| - (1-\gamma)\omega_i)\right)\right], \text{ if } (1-\gamma)\omega_i \leq |\omega| \leq (1+\gamma)\omega_i \\ 0, \text{ Otherwise} \end{cases} \tag{16}$$

and

$$\varphi_i(\omega) = \begin{cases} 1, \text{ if } |\omega| \leq (1-\gamma)\omega_i \\ \cos\left[\frac{\pi}{2}\beta\left(\frac{1}{2\gamma\omega_i}(|\omega| - (1-\gamma)\omega_i)\right)\right], \text{ if } (1-\gamma)\omega_i \leq |\omega| \leq (1+\gamma)\omega_i \\ 0, \text{ Otherwise} \end{cases} \tag{17}$$

where $\gamma = min_i\left(\frac{\omega_{i+1}-\omega_i}{\omega_{i+1}+\omega_i}\right)$. In both the above equations, the function $\beta(x)$ represents an arbitrary $C^k([0,1])$ function, such that,

$$\beta(x) = \begin{cases} 1\,if\,x \geq 1 \\ 0\,if\,x \leq 0 \end{cases} \text{ and } \beta(x) + \beta(1-x) = 1 \;\; \forall x \in [0,1] \tag{18}$$

Most of the functions do satisfy these properties, and studies such as Gilles [22] and Daubechies [34] have shown that the frequently used is as follows:

$$\beta(x) = x^4\left(35 - 84x + 70x^2 - 20x^4\right) \tag{19}$$

It should be noted that the above equations were reduced to this form after first assuming that the Fourier support $[0,\pi]$ is segmented into N contiguous segments (e.g., Gilles [22]). The frequency ω_i is then denoted to be the limits between each segment. A detailed procedure followed to produce both the empirical wavelet and empirical scaling function is given in a study by Gilles [22].

The fourth step is to calculate the approximate and detail coefficients using the following equations:

$$W_x(i,t) = \langle x(t), \psi_i(t)\rangle = \int x(\tau)\psi_i(\tau - t)d = F^{-1}[x(\omega)\psi(\omega)] \tag{20}$$

$$W_x(1,t) = \langle x(t), \psi_1(t)\rangle = \int x(\tau)\psi_1(\tau - t)d = F^{-1}[x(\omega)\psi_i(\omega) \tag{21}$$

The final step is to reconstruct the original signal in order to obtain different modes in the following form:

$$x(t) = W_x(1,t) \times \varphi_1(t) + \sum_{i=2}^{N} W_x(i,t) \times \psi_i(t) \tag{22}$$

The details about the EWT contraction process as well as the testing of the method using different signals, including artificial signals and real electrocardiogram signals, is decsribed in Gilles [22]. Recently, Zhou et al. [17] used the EWT technique in a data pre-analysis step in their study of data pre-analysis and ensemble of various artificial neural networks for monthly streamflow forecasting.

2.5. *Long Short-Term Memory (LSTM)*

The LSTM is a special type of recurrent neural network (RNN) that was proposed by Hochreiter and Shmidhuber [23]. These RNNs are improved multilayer perception networks that are often termed as deep learning, and are somewhat different from traditional artificial neural networks (ANN) [35]. The general architecture of the traditional ANNs employs a feedforward neural network system, which means artificial neural networks with connections between the units do not form a cycle. In other words, in a feedforward network, the processing of information is piped through the network from the input layers to output layers. On the other hand, the RNN is an artificial neural network where connections form between units form cyclic paths. RNNs are defined as recurrent because, in their architecture, they (1) receive input, (2) update the hidden states depending on the previous computations, and (3) make predictions for every element of a sequence. Therefore, these types of neural networks are considered to have memory because they keep information about what has been processed. More details about RNNs can be found in a study by Hochreiter and Shmidhuber [23]. The architecture of the RNN is summarised in Figure 1.

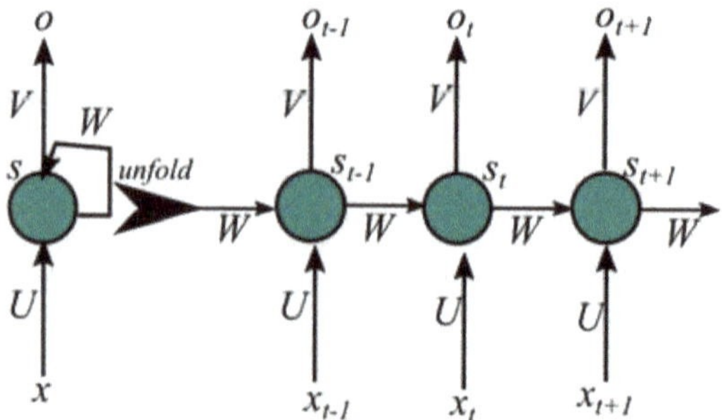

Figure 1. Architecture of a recurrent neural network (RNN).

In this figure (Figure 1), x_t represents the input time series at time step t. The RNN computes the hidden state sequence s_t at given time step t as well as the output sequence $y = \{y_1, y_2, \ldots, y_t\}$. Unlike the traditional ANN, the RNN shares the same network parameters (U, V, W) across all steps. The network illustrated in Figure 1 can also be mathematically represented as follows:

$$s_t = f(Ux_t + WS_{t-1}) \tag{23}$$

$$y = g(Vs_t) \tag{24}$$

where parameters g and f are the active functions for the output layer and hidden layer, respectively.

As mentioned above, LSTMs are a family of RNNs that are often used with deep neural networks. In summary, and as illustrated by Figure 2, an LSTM is made up of three gates: (1) a forget gate (*f_t*), which controls if/when the context is forgotten; (2) an input gate (*i_t*), which controls if/when a value should be remembered by the context; and (3) an output gate (*o_t*), which controls if/when the remembered value is allowed to pass from the unit. This exclusive structure of the LSTM is capable of effectively solving the problem of gradient loss and gradient explosion problems during the training procedure [23].

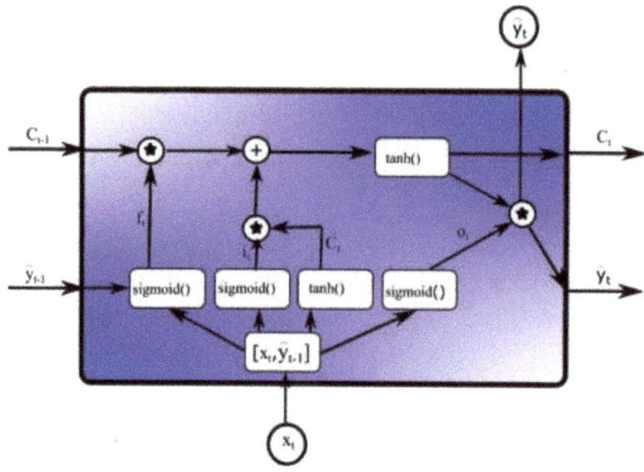

Figure 2. The structure of the long short-term memory (LSTM) unit.

In a mathematical form, the above diagram (Figure 2) can be represented as follows:

$$f_t = S\left(W_f \cdot [\hat{y}_{t-1}, x_t] + b_f\right) \tag{25}$$

$$i_t = S(W_i \cdot [\hat{y}_{t-1}, x_t] + b_i) \tag{26}$$

$$\widetilde{C}_t = tanh(W_C \cdot [\hat{y}_{t-1}, x_t] + b_C) \tag{27}$$

$$C = f_t \cdot C_{t-1} + i_t \cdot \hat{C}_t \tag{28}$$

$$o_t = S(W_o \cdot [\hat{y}_{t-1}, x_t] + b_o) \tag{29}$$

$$\hat{y}_t = o_t \cdot tanh(C_t) \tag{30}$$

2.6. Novel Hybrid Model Design

The use of the EWT to decompose data before use in the model is expected to improve the prediction accuracy and to solve the issue of long-term dependencies forecasting problems of the TCO monthly time series, such as the one used here. The EWT-LSTM developed here is also compared to the hybrid method based on the EMD and LSTM neural networks, namely EMD-LSTM. This hybrid model (EMD-LSTM) was applied to geophysical data for the first time by Zhang et al. [11]. The two above models are also compared to the LSTM model that uses wavelet denoised (WD) data (WD-LSTM), wavelet denoising EWT-LSTM (WD-EWT-LSTM), and wavelet denoising EMD-LSTM (WD-EMD-LSTM). For denoising, this study uses a Daubechies (db8) wavelet family in the PyWavelets Wavelet Transform software for Python. This family of wavelets is selected because it performed better than others in terms of applications in the forecasting of the TCO data. In addition, the LSTM system used here is a Python TensorFlow LSTM system that is run in miniconda [36].

Figure 3 illustrates the detailed procedures of the architecture of the hybrid model used in this study. Apart from the models that used the Wavelet denoising step, implementation of the EWT-LSTM and EMD-LSTM models developed here occurs in three important steps. The first step utilizes the EMD or EWT technique to decompose the TCO time series data into relatively stable IMFs and one residue item. In the second step, the LSTM neural network is used to forecast each normalized IMF, including the residue. In the third step, the forecasting results of the LSTM neural network are reverse-normalized before being accumulated through summation as the final predicted results. It is important to note that the TCO data were divided into 70% training and 30% testing datasets. The training dataset is used for the LSTM modeling, and the testing dataset is used as an input into the trained LSTM models to predict all the IMFs and the residue. After obtaining the final forecasting values, model performance evaluation is applied, utilizing several statistical metrics to assess the performance of all six models presented here.

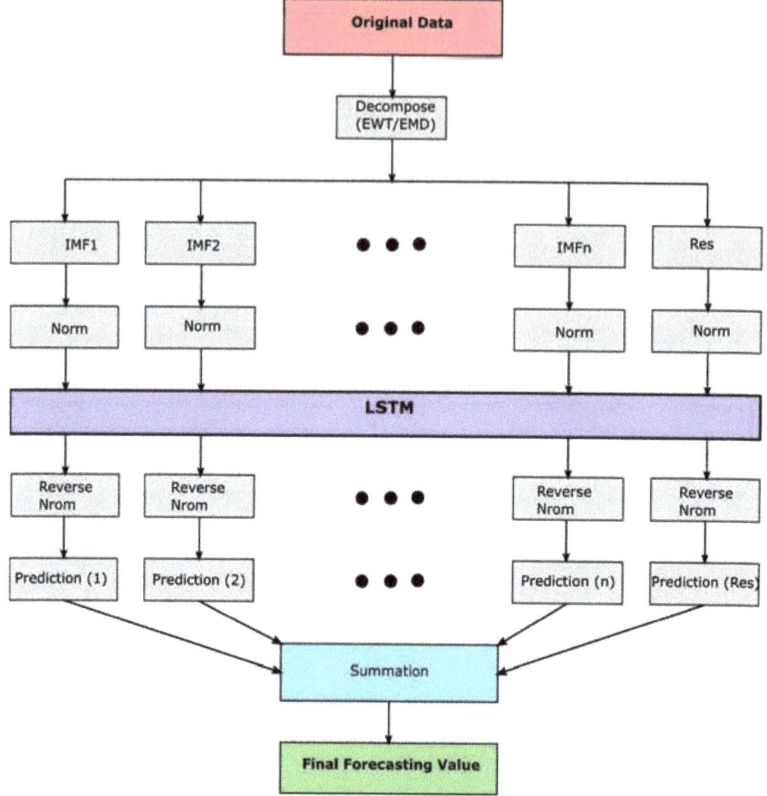

Figure 3. The architecture of the data driven LSTM hybrid model. IMF, intrinsic mode function; EWT, empirical wavelet transform; EMD, empirical mode decomposition.

2.7. Model Performance

In general, there is no standard criterion for assessing the forecasting performance of a model and comparison with other benchmark models [37]. In this study, the model performance evaluation is done by comparing the predicted values with their corresponding observed values using typical performance metrics, which are summarized in Table 1.

Table 1. Model performance metrics.

Metric	Definition	Formula
MAE	Mean absolute error	$MAE = \frac{1}{N}\sum_{j=1}^{N}\left\lvert p^{j}_{estimated} - p^{j}_{real}\right\rvert$
RMSE	Root mean square error	$RMSE = \sqrt{\frac{1}{N}\sum_{j=1}^{N}\left(p^{j}_{estimated} - p^{j}_{real}\right)^2}$
MAPE	Mean absolute percentage error	$MAPE = \frac{1}{N}\sum_{j=1}^{N}\left\lvert\frac{p^{j}_{estimated} - p^{j}_{real}}{p^{j}_{real}}\right\rvert$

3. Results and Discussion

3.1. TCO Data Series and Trends

Figure 4a shows the monthly means of TCO time series measured at Observatorio Central Buenos Aires (OCBA) during the period 1966–2017. In general, the TCO time series over the study area indicates a strong seasonal variability, with maximum ozone during summer season and minimum values during winter.

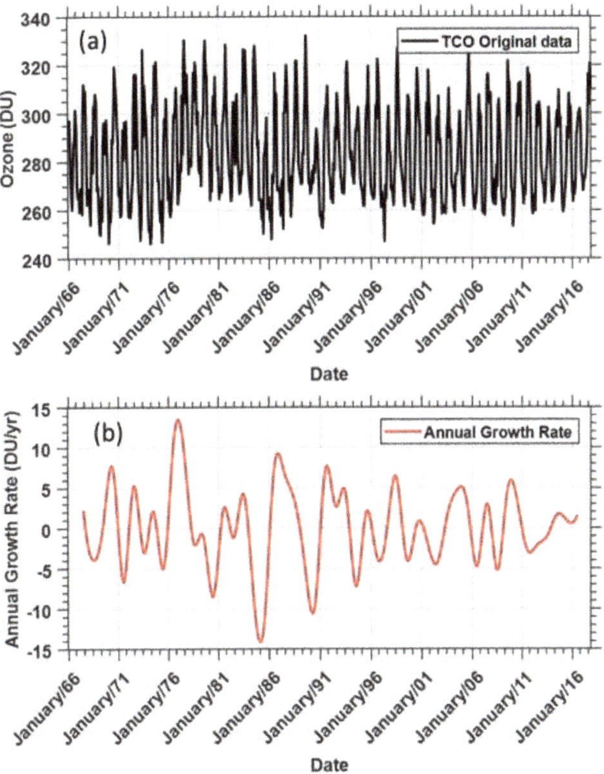

Figure 4. Total columns of ozone (TCO) time series at the Observatorio Central Buenos Aires (OCBA) (**a**) and its corresponding annual mean growth rate (**b**).

Owing to the dynamics of the TCO especially in the Southern Hemisphere, it is also important to consider calculating the annual growth rate of the TCO data used here. This growth rate is defined as the instantaneous slope of the de-seasonalised curve. In this study, the annual growth rate was computed with the algorithm used by NOAA (National Oceanic and Atmosphere Administration), which was developed by Thoning et al. [38]. This method has been shown to be useful in extracting annual growth rates in a number of atmospheric chemistry climatology studies, including a recent study by Apadula et al. [39]. The method uses monthly mean TCO values corrected for the seasonal cycles, computes the averages of the four November, December, January, and February months every year, and associates these values to 1 January. The estimated value for the annual mean growth rate is then computed by evaluating the differences between the averages. Our study utilizes a twelve-month centered moving mean value for the OCBA TCO time series. The annual growth rate for the OCBA TCO time series is shown in Figure 4b.

In general, the mean growth rate for the OCBA TCO time series for the period from 1966 to 2017 is equal to 0.08 DU/year. An inspection of Figure 4b shows that the highest growth rate in the OCBA TCO data is obtained for the years 1976–1977 (13.5 DU/year) followed by 1988–1989 (8.6 DU/year). On the other hand, the lowest annual growth rate value was observed in 1987 (−14.1 DU/year). Overall, there was a strong variability of the TCO growth rate during the period between 1974 and 1984. The observed changes in the annual growth rate variability during the mid-nineties could be associated with the early successes of the Montreal Protocol implementation strategies, banning substances that depleted ozone. Moreover, this takes place at a similar time to when the upper stratospheric HCL, a proxy for stratospheric chlorine, started to decline from the early 2000s, as reported by Fioletov [40]. However, during this period, the annual growth rate recorded values above 5 DU/year during the 1997–1998, 2003, and 2009–2010 El Niño years, which is more likely the cause, As the El Niño-Southern Oscillation (ENSO) is known to influence TCO [41]. The National Aeronautics and Space Administration Ozone Watch [42] reported the year 1987 as one of the years that showed the greatest rate of ozone decrease as well as the longest persistence of the ozone hole in Southern Hemisphere. The growth rate time series presented in Figure 4b seems to be consistent with these findings because it experiences its highest minimum (−14 DU/year) during the year 1987. The observed growth rate minimas seem to coincide with ozone depletion events, resulting from some of the largest volcanic eruptions in the El Chichón in 1982 and Mount Pinatubo in 1991.

In this study, two powerful methods for conducting the decomposition of the non-stationary and non-linear signal are used, namely EMD and EWT. The decomposition of the time series is an integral part of the hybrid model for forecasting the TCO time series proposed here. The EMD method decomposed the original data series into seven relatively stable IMFs and one residue item (Figure 5). On the other hand, the EWT method was also applied to the original data series, and it produced twelve IMFs and one residue item (Figure 6). Generally, the IMFs extracted by both the EMD and EWT methods, indicate (1) the characteristics of lower and middle atmosphere time series data where the starting IMFs represent lower frequency and (2) the latter half IMFs indicate high frequencies, and (3) residuals are shown as trends. These characteristics indicate the presence of various periodicities such as solar cycle, QBO, annual cycle, and semi-annual cycle, among other periodicities. For the purpose of this study, the EMD amplitude of white noise was set to 0.2, and the ensemble number was set to 1000. In addition, all sub-band signals extracted by EMD and EWT were used in the decomposition step of the hybrid LSTM model.

Generally, the limited availability of methods that have well-defined theory that can assist in the extraction of a trend in a time series poses a problem. The singular spectrum analysis (SSA) [43] trend extraction technique is a well-known method that captures the trend of a signal via eigenvalues from a trajectory matrix. Moreover, the EMD data-driven technique has been shown to offer the possibility of estimating a trend by summation of low-frequency intrinsic mode functions [44]. Previous studies have shown that EWT has an advantage over EMD, in addition to having a well-defined theory [22,45]. In our study, we successfully extracted the trend of the TCO time series (shown in Figure 4a) by means of the three above-mentioned methods (Figure 7). The EWT trend used here is the residue of the EWT, while the EMD trend was extracted via the summation of low-frequency intrinsic mode functions produced by the EMD [44]. Regarding the EMD, it is important to investigate whether each IMF (including the trend) is composed of useful information or just noise. Therefore, in this paper, we followed a method outlined by Wu and Huang [46], and also utilized features applied by Sang et al. [47] in their study of a comparison of the Mann-Kendal test and EMD method for trend identification in hydrological time series, in order to assess the statistical significance of EMD IMFs. These two trend analysis methods (EWT and EMD) are compared with the SSA trend method (Figure 7).

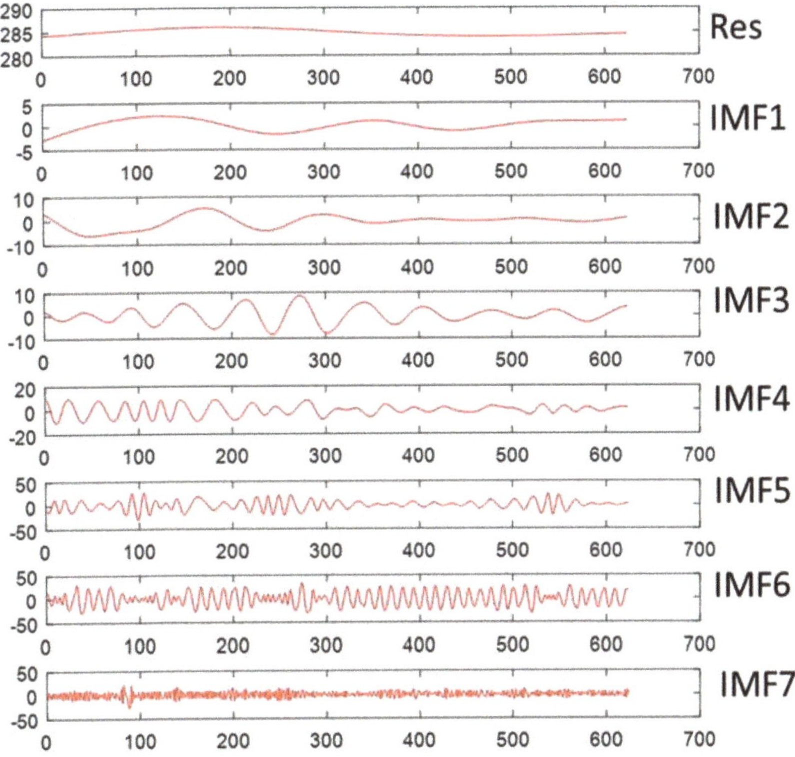

Figure 5. Modes extracted by empirical mode decomposition (EMD) for OCBA TOC ozone time series.

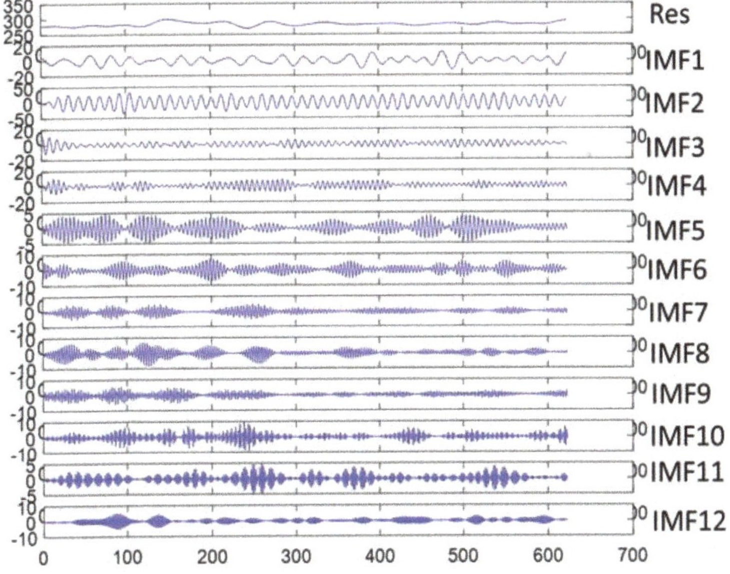

Figure 6. Modes extracted by empirical wavelet transform (EWT) for OCBA TCO ozone time series.

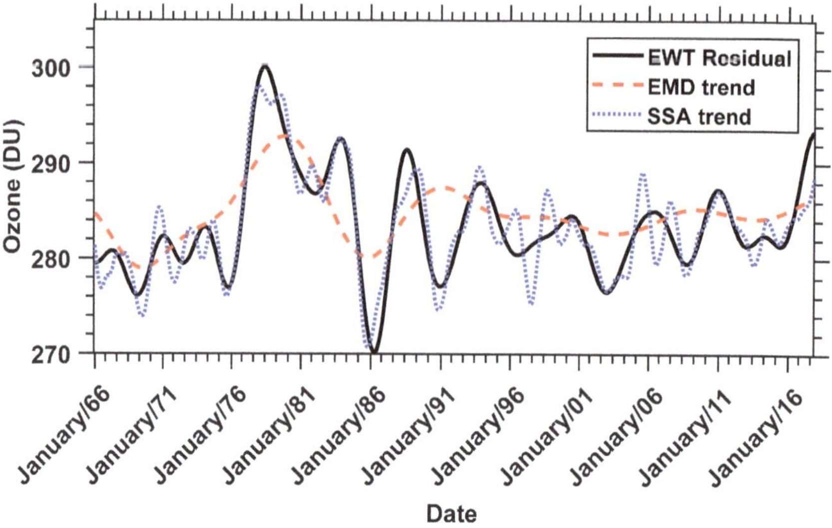

Figure 7. OCBA TCO trend time series extracted by (EWT) (black line), EMD (red dashed line), and singular spectrum analysis (SSA) (blue dotted line).

In Figure 7, it is apparent that the EWT residue is able to mimic the SSA trend very well. However, the SSA trend method is influenced by the 32- to 64-month periodicity, which is observed to be smoothened out in both the EWT and the EMD methods. Overall, the three methods used here are able to capture the disturbances of the TCO over Buenos Aires, which are associated with major events such as the 1987 major decline of ozone in the South Pole, as well as during some of the largest volcanic eruptions during the 1982 El Chichón and Mount Pinatubo during 1991. The observed TCO trend declines from the early 1980s before it becomes steady from the early 2000s onwards. The turning of the trend to an upwards direction is observed later in the time series, which indicates the effect of the Montreal Protocol. The Montreal Protocol and its subsequent amendments are known to have successfully prevented catastrophic losses of the stratospheric ozone [9].

In this study, we utilized several time series analysis techniques such as wavelet analysis, Mann-Kendall test, and sequential Mann-Kendall test model, and applied them to the Buenos Aires TCO data. The wavelet analysis technique was efficient in analyzing the localized variation of the spectral power within the OCBA TCO data. Overall, wavelet analysis is a common tool for this purpose and this signal processing method makes it possible to determine the dominant modes of variability, in addition to their variation within the time series. The annual Mann-Kendall trend test method was also performed by calculating annual z-score values throughout the time series. We found this non-parametric trend test method very suitable for the TCO trend analysis for two reasons: (1) it does not require the data to be normally distributed, and (2) it does not depend on the magnitude of the data. Furthermore, this method was supported by employing the sequential version of the Mann-Kendall test, in order to determine the change detection points in the OCBA TCO time series trend. Further details about the theory of the above mentioned time series analysis methods and their application can be found in a study by Mbatha et al. [28].

Figure 8 shows OCBA TCO monthly mean time series for the period from 1966 to 2017 with its smooth trend (a), normalized wavelet power spectra of the TCO data (b), annual mean Mann-Kendall calculated for TCO time series (c), and sequential Mann-Kendall statistics of progressive (Prog) $u(t)$ and retrograde (Retr) $u'(t)$ (d). In the normalized wavelet power spectra in Figure 8b, the white thick lines represent the 95% confidence level, and areas of the wavelet power that are considered are those that are within the cone-of-influence (indicated by solid upside-down "u" shaped line). As expected, a strong peak at around the 12-month cycle dominates the wavelet power spectra of Buenos Aires TCO monthly mean time series. Apart from the dominant annual oscillation, we observe that there are prominent oscillations presented in the TCO time series, having periodicities of 16 to 18 months, 28 to 32 months (QBO periodicity), quasi 64 months, and evidence of a strong 11-year solar cycle.

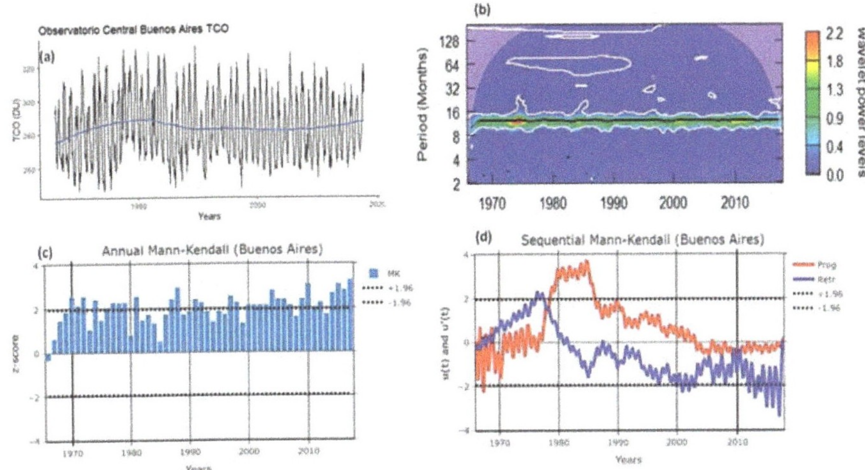

Figure 8. OCBA TCO time series with its smooth trend (**a**), normalized wavelet power spectra of the TCO data (**b**), annual mean Mann-Kendall calculated for TCO (**c**), and sequential Mann-Kendall statistics values (**d**).

The annual Mann-Kendall test time series shown in Figure 8c indicates distinctive periods of positive and significant z-score (greater than $Z1$ = 1.96), especially during major events, which are summarized in the TCO growth rate analysis in Figure 4. The positive significant z-score values seem to dominate in the period after the year 2000, with a similar period from 2014 to 2017, experiencing consistent positive z-score values. This is indicative of the effect of the Montreal Protocol agreement, and hence signs of the recovery of stratospheric ozone. This is also in agreement with the trends reported by SQ-MK in Figure 8d, indicating a downward trend for both the forward and backward (Prog and Retr) statistic values, having started in the 1970s and 1980s. Stabilization of this downward trend is observed from the beginning of the 2000s and later takes an upward (but not significant) direction in subsequent years. The specific change detection point is in the mid-2000s and during 2010.

3.2. *Empirical Models Results and Its Performance*

To improve the prediction accuracy of complex geophysical data from a simple time series model, one needs to consider taking advantage of signal preprocessing methods such as denoising, decomposition, and ensembling the predicted results. Therefore, to understand the performance of the hybrid EWT-LSTM model, the predicted results of the EWT-LSTM model are compared with the LSTM, EMD-LSTM, WD-LSTM, WD-EMD-LSTM, and WD-EWT-LSTM. In this study, the OCBA TCO time series is divided into 70% training part of the time series and the 30% testing part of the time series. Thus, the model was trained using data from January 1966 to May 2002, and the model forecast testing part of the time series started from 2002 to 2017. Figure 9 shows the predicted results of the six models plotted together with the original time series (testing time series). In this figure, it is apparent that the six models predict different forecast results of the TCO time series measured at the OCBA site. However, the hybrid EWT-LSTM, WD-EWT-LSTM, and EMD-LSTM models seem to perform better predictions compared with other ones. While it is difficult to comment on the other models, the single LSTM model is observed to hardly catch the sudden changes in the original time series.

In order to further assess the prediction performance of the six models used here, a Taylor diagram [48] and statistical evaluation metrics such as MAE, MAPE, R, RMSE, and STD are utilized to measure the performance the models. The Taylor diagram is informative as it assists by visualization of the comparative strength of the models to the actual target variable. In the Taylor diagram, two different statistical metrics (i.e., correlation coefficients "R" and standard deviations "STD" of each model) are used to quantifying the comparability between the models and the observational data. The distance from the reference point, which is the observed data, is a measure of the cantered RMSE.

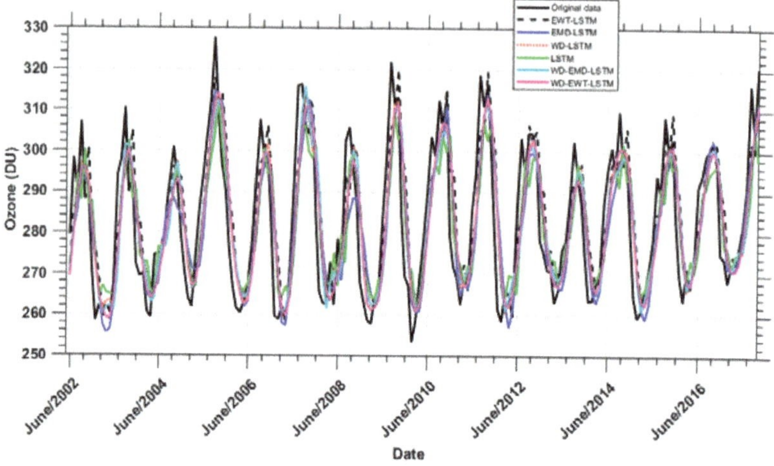

Figure 9. Forecasting results of the OCBA TCO time series for the period of June 2002 to December 2017 as derived by the use of six models (LSTM, EWT-LSTM, EMD-LSTM, wavelet denoising (WD)-LSTM, WD-EWT-LSTM, and WD-EMD-LSTM) together with the TCO monthly mean values derived from observations (original data).

Figure 10 shows the Taylor diagram of the models used in this study. In this figure, owing to the denseness of the models, a zoomed view of the models position in the Taylor diagram is shown in the right panel of Figure 10. On the basis of the representation in Figure 10 for the six data-driven models used in this study, EWT-LSTM outperformed the other five models. The EWT-LSTM is the closest to the observed point, which means it has the smallest RMSE. EWT-LSTM also has the highest correlation of approximately R = 0.87, and the smallest standard deviation. The Taylor diagram also shows that time series preprocessing methods such as denoising and decomposition play a significant role in improving the accuracy of the model. This is because the LSTM model, a model that does not use preprocessed data, is the worst performing model in the view of the Taylor diagram.

Figure 11 shows a bar graph that summaries the comparison of the model forecasting performance evaluated using three criteria, namely, RMSE, MAE, and MAPE. A visual inspection of Figure 9 and the Taylor diagram in Figure 10 seems to indicate the existence of a preliminary judgment that the forecasting accuracy of the EWT-LSTM model is higher than that of the other five models proposed in this study. Moreover, it can be observed in Figure 11 that, among all the forecasting models used here, EWT-LSTM performs better than any other model, namely, LSTM, WD-EWT-LSTM, EMD, and WD-EMD-LSTM, in terms of RMSE, MAE, and MAPE, which are 9.4, 7.5, and 2.6%, respectively. The LSTM model, a model that does not consider any of the data preprocessing methods used in this study, is the worst performing model. The aforementioned is presumably owing to the non-stationary and non-linear nature of the original monthly mean TCO data series, which is captured by the preprocessing methods used here. On the other hand, preprocessing of data with methods such as denoising and decomposition is known to improve the performance of the LSTM model and other artificial neural network models [11,14,17]. The use of the wavelet transform denoising technique, a model design reported for the first time in this study, seems to improve the LSTM model accuracy. Better model accuracy is obtained when the model's original data are decomposed using EMD [11] or EWT. However, denoising the data before decomposing and then applying the LSTM neural network on it does not improve the model. This is because LSTM is able to learn both long-term and short-term variation in the data, which also accommodates some level of noise in the data. Moreover, it is possible that the denoising step also removes some important pattern in the data across time. Generally, EWT, a new method to detect the principal "modes" that represent the signal, and with good mathematical theory, seems to perform better than the popular EMD method, which lacks mathematical theory when applied in the decomposition step of the model.

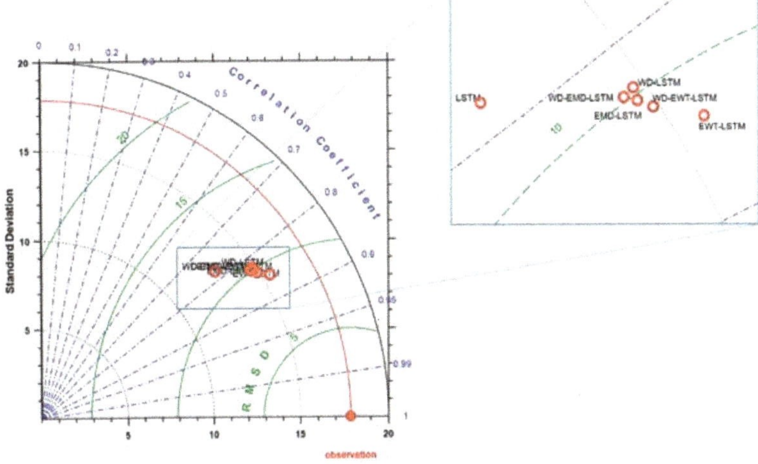

Figure 10. Taylor diagram graphical representation of six predictive models developed for forecasting OCBA TCO time series from 2002 to 2017.

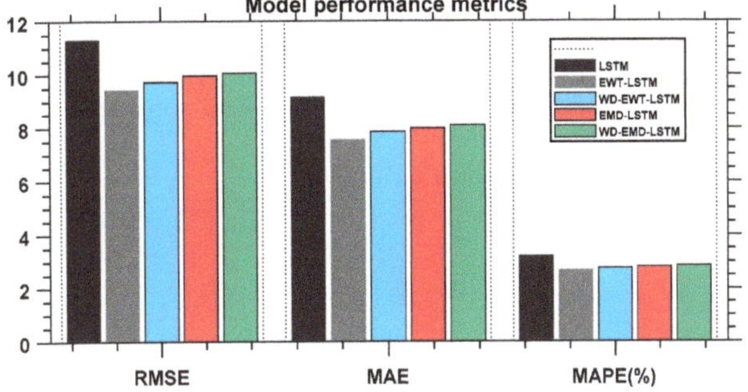

Figure 11. Error graphics of models performance of OCBA TCO time series. RMSE, root mean square error; MAE, mean absolute error; MAPE, mean absolute percentage error.

4. Conclusions

The present study utilized different time series analysis methods to investigate the trend and variability of the TCO monthly mean time series, in addition to developing and testing six LSTM recurrent neural networks based data-driven time series forecasting models. The TCO data, long-term ozone data measured at the Buenos Aires site, Argentina, from 1966 to 2017, were used in this study.

Overall, the trends of TCO time series data extracted using the SSA, EWT, EMD, and Mann-Kendall confirm that the TCO has been stable since the mid-1990s. The SSA growth rate and trend, EMD trend, and EWT trend capture the dynamical variability of the TCO well, and their results are somewhat consistent. All trend analysis methods seem to report a significant recovery of ozone during the period from 2010 to 2017, apart from the decline of ozone observed during 2015, which is presumably associated with the Calbuco volcanic event (a Chilean volcano that has injected volcanic plume up to the stratosphere, as reported by Bègue et al. [49,50]). The sensitivity of the EWT trend results presented in Figure 7 seems to indicate that the EWT extraction method, a method that has a well-defined mathematical theory compared with others, can be the best method for trend extraction.

Six LSTM neural networks based data-driven time series forecasting models, namely, LSTM, EWT-LSTM, EMD-LSTM, WD-LSTM, WD-EWT-LSTM, and WD-EMD-LSTM, were developed in this research. The mode that used the EWT decomposition process (EWT-LSTM) outperformed the other five models in terms of forecasting performance evaluation criteria, such as the RMSE, MAE, MAPE, and R. In general, better model accuracy was achieved for those models that used decomposition methods, demonstrating that EMD and EWT methods play a significant role in the improvement of the performance of the LSTM model. Noise reduction techniques via wavelet transform improved the LSTM model accuracy, but it appears unnecessary to denoise data to the data when utilizing the decomposition process. Overall, the results presented in this study show that the EWT-LSTM model can be used as a successful tool for monthly mean TCO forecasting.

For future work, it is recommended to use a method that has good mathematical theory such as the EWT technique to perform trend analysis of ozone data, or any other geophysical time series. The continuation of this study will include applying the techniques developed to the stratospheric column ozone, lower stratosphere ozone, and upper stratosphere ozone measured in the high latitude Southern Hemisphere, in order to forecast the process of ozone hole recovery. Additionally, more TCO measuring sites at different latitudes will also be considered in future studies.

Author Contributions: Conceptualization, N.M. and H.B.; methodology, N.M.; software, N.M.; validation, N.M.; formal analysis, N.M.; investigation, N.M. and H.B.; writing—original draft preparation, N.M. and H.B.; writing—review and editing, N.M. and H.B.; visualization, N.M. All authors have read and agreed to the published version of the manuscript.

Funding: This research was funded jointly by the CNRS (Centre National de la Recherche Scientifique) and the NRF (National Research Foundation) in the framework of the LIA ARSAIO and by the South Africa/France PROTEA Program (project No 42470VA).

Acknowledgments: Authors acknowledge the French South-African PROTEA programme and the CNRS-NRF LIA ARSAIO (Atmospheric Research in Southern Africa and Indian Ocean), for supporting research activities, and the National Research Foundation (NRF) of South Africa. We are thankful to the World Ozone and Ultraviolet Radiation Data Centre (WOUDC) for data archiving, the PI and operators of Dobson instrument at Buenos Aires, and the National Meteorological Service of Argentina (SMNA).

Conflicts of Interest: The authors declare no conflict of interest.

References

1. Weber, M.; Coldewey-Egbers, M.; Fioletov, V.E.; Frith, S.M.; Wild, J.D.; Burrows, J.P.; Long, C.S.; Loyola, D. Total ozone trends from 1979 to 2016 derived from five merged observational datasets – the emergence into ozone recovery. *Atmos. Chem. Phys.* **2018**, *18*, 2097–2117. [CrossRef]
2. Farman, J.C.; Gardiner, B.G.; Shanklin, J.D. Large losses of total ozone in Antarctica reveal seasonal ClO x/NO x interaction. *Nature* **1985**, *315*, 207–210. [CrossRef]
3. Solomon, S.; Garcia, R.R.; Rowland, F.S.; Wuebbles, D.J. On the depletion of Antarctic ozone. *Nature* **1986**, *321*, 755–758. [CrossRef]
4. Rigby, M.; Park, S.; Saito, T.; Western, L.M.; Redington, A.L.; Fang, X.; Henne, S.; Manning, A.J.; Prinn, R.G.; Dutton, G.S. Increase in CFC-11 emissions from eastern China based on atmospheric observations. *Nature* **2019**, *569*, 546–550. [CrossRef] [PubMed]

5. Ball, W.T.; Alsing, J.; Mortlock, D.J.; Staehelin, J.; Haigh, J.D.; Peter, T.; Tummon, F.; Stübi, R.; Stenke, A.; Anderson, J.; et al. Evidence for a continuous decline in lower stratospheric ozone offsetting ozone layer recovery. *Atmos. Chem. Phys.* **2018**, *18*, 1379–1394. [CrossRef]
6. Braesicke, P.; Neu, J.; Fioletov, V.; Godin-Beekmann, S.; Hubert, D.; Petropavlovskikh, I.; Shiotani, M.; Sinnhuber, B.-M. Update on Global Ozone: Past, Present, and Future. In *Scientific Assessment of Ozone Depletion: 2018*; Global Ozone Research and Monitoring Project–Report No. 58; World Meteorological Organization: Geneva, Switzerland, 2018; Chapter 3.
7. Pawson, S.; Steinbrecht, W.; Charlton-Perez, A.J.; Fujiwara, M.; Karpechko, A.Y.; Petropavlovskikh, I.; Urban, J.; Weber, M.; Aquila, V.; Chehade, W. Update on global ozone: Past, present, and future. In *Scientific Assessment of Ozone Depletion: 2014*; Global Ozone Research and Monitoring Project–Report No. 55; World Meteorological Organization: Geneva, Switzerland, 2014; Chapter 3.
8. Chehade, W.; Weber, M.; Burrows, J.P. Total ozone trends and variability during 1979–2012 from merged data sets of various satellites. *Atmos. Chem. Phys.* **2014**, *14*, 7059–7074. [CrossRef]
9. Ball, W.T.; Alsing, J.; Staehelin, J.; Davis, S.M.; Froidevaux, L.; Peter, T. Stratospheric ozone trends for 1985–2018: Sensitivity to recent large variability. *Atmos. Chem. Phys.* **2019**, *19*, 12731–12748. [CrossRef]
10. Lei, M.; Shiyan, L.; Chuanwen, J.; Hongling, L.; Yan, Z. A review on the forecasting of wind speed and generated power. *Renew. Sust. Energ. Rev.* **2009**, *13*, 915–920. [CrossRef]
11. Zhang, X.; Zhang, Q.; Zhang, G.; Nie, Z.; Gui, Z.; Que, H. A Novel Hybrid Data-Driven Model for Daily Land Surface Temperature Forecasting Using Long Short-Term Memory Neural Network Based on Ensemble Empirical Mode Decomposition. *Int. J. Environ. Res. Public. Health.* **2018**, *15*, 1032. [CrossRef]
12. Zhang, G.P.; Qi, M. Neural network forecasting for seasonal and trend time series. *Eur. J. Oper. Res.* **2005**, *160*, 501–514. [CrossRef]
13. Tealab, A. Time series forecasting using artificial neural networks methodologies: A systematic review. *future computing inform. j.* **2018**, *3*, 334–340. [CrossRef]
14. Tian, C.; Hao, Y.; Hu, J. A novel wind speed forecasting system based on hybrid data preprocessing and multi-objective optimization. *Applied Energy* **2018**, *231*, 301–319. [CrossRef]
15. Zhang, G.P. Time series forecasting using a hybrid ARIMA and neural network model. *Neurocomputing* **2003**, *50*, 159–175. [CrossRef]
16. Khandelwal, I.; Adhikari, R.; Verma, G. Time Series Forecasting Using Hybrid ARIMA and ANN Models Based on DWT Decomposition. *Procedia Comput. Sci.* **2015**, *48*, 173–179. [CrossRef]
17. Zhou, J.; Peng, T.; Zhang, C.; Sun, N. Data Pre-Analysis and Ensemble of Various Artificial Neural Networks for Monthly Streamflow Forecasting. *Water* **2018**, *10*, 628. [CrossRef]
18. Altan, A.; Karasu, S.; Bekiros, S. Digital currency forecasting with chaotic meta-heuristic bio-inspired signal processing techniques. *Chaos Soliton. Fract.* **2019**, *126*, 325–336. [CrossRef]
19. Liu, Y.; Guan, L.; Hou, C.; Han, H.; Liu, Z.; Sun, Y.; Zheng, M. Wind Power Short-Term Prediction Based on LSTM and Discrete Wavelet Transform. *Appl. Sci.* **2019**, *9*, 1108. [CrossRef]
20. Li, Y.; Wu, H.; Liu, H. Multi-step wind speed forecasting using EWT decomposition, LSTM principal computing, RELM subordinate computing and IEWT reconstruction. *Energy Convers. Manag.* **2018**, *167*, 203–219. [CrossRef]
21. Nazir, H.M.; Hussain, I.; Faisal, M.; Shoukry, A.M.; Gani, S.; Ahmad, I. Development of Multidecomposition Hybrid Model for Hydrological Time Series Analysis. *Complexity* **2019**, *2019*, 1–14. [CrossRef]
22. Gilles, J. Empirical wavelet transform. *IEEE Trans. Signal. Process.* **2013**, *61*, 3999–4010. [CrossRef]
23. Hochreiter, S.; Schmidhuber, J. Long Short-Term Memory. *Neural Computation* **1997**, *9*, 1735–1780. [CrossRef] [PubMed]
24. World Ozone and Ultraviolet Radiation Data Centre website. Available online: https://woudc.org/home.php (accessed on 30 June 2018).
25. Dobson, G.M.B.; Harrison, D.N.; Lindemann, F.A. Measurements of the amount of ozone in the earth's atmosphere and its relation to other geophysical conditions. *P. R. Soc. A-Math. Phy.* **1926**, *110*, 660–693. [CrossRef]
26. Stolarki, R.S.; Krueger, A.J.; Schoeberl, M.R.; McPeters, R.D.; Newman, P.A.; Alpert, J.C. Nimbus 7 SBUV/TOMS measurements of the springtime Antarctic ozone hole. *Nature* **1986**, *322*, 808–811. [CrossRef]
27. Grant, W.B. *Ozone Measuring Instruments for the Stratosphere*; Collection Work in Optics; Optical Society of Amer: Washington DC, USA, 1989.

28. Mbatha, N.; Xulu, S. Time Series Analysis of MODIS-Derived NDVI for the Hluhluwe-Imfolozi Park, South Africa: Impact of Recent Intense Drought. *Climate* **2018**, *6*, 95. [CrossRef]
29. Mann, H.B. Nonparametric tests against trend. *Econometrica.* **1945**, *13*, 245–259. [CrossRef]
30. Sneyers, R. *On the statistical analysis of series of observations*; Technical Note 143; World Metrological Organization, : Geneva, Switzerland, 1991.
31. Pohlert, T. Trend: Non-Parametric Trend Tests and Change-Point Detection. Available online: https://cran.r-project.org/web/packages/trend/index.html (accessed on 6 August 2018).
32. Schaber, J. Pheno: Auxiliary Functions for Phenological Data Analysis. Available online: https://cran.r-project.org/web/packages/pheno/index.html (accessed on 6 August 2018).
33. Torres, M.E.; Colominas, M.A.; Schlotthauer, G.; Flandrin, P. A complete ensemble empirical mode decomposition with adaptive noise. In Proceedings of the 2011 IEEE International Conference on Acoustics, Speech and Signal Processing (ICASSP), Prague, Czech Republic, 22–27 May 2011; pp. 4144–4147.
34. Daubechies, I. *Ten Lectures on Wavelets*; SIAM: Bangkok, Thailand, 1992.
35. Giles, C.L.; Lawrence, S.; Tsoi, A.C. Noisy Time Series Prediction using Recurrent Neural Networks and Grammatical Inference. *Mach. Learn.* **2001**, *44*, 161–183. [CrossRef]
36. Miniconda—Conda documentation. Available online: https://docs.conda.io/en/latest/miniconda.html (accessed on 1 May 2019).
37. Xu, Y.; Yang, W.; Wang, J. Air quality early-warning system for cities in China. *Atmos. Environ.* **2017**, *148*, 239–257. [CrossRef]
38. Thoning, K.W.; Tans, P.P.; Komhyr, W.D. Atmospheric carbon dioxide at Mauna Loa Observatory: 2. Analysis of the NOAA GMCC data, 1974–1985. *J. Geophys. Res: Atmos.* **1989**, *94*, 8549–8565. [CrossRef]
39. Apadula, F.; Cassardo, C.; Ferrarese, S.; Heltai, D.; Lanza, A. Thirty Years of Atmospheric CO2 Observations at the Plateau Rosa Station, Italy. *Atmosphere* **2019**, *10*, 418. [CrossRef]
40. Fioletov, V.E. Ozone climatology, trends, and substances that control ozone. *Atmos. Ocean.* **2008**, *46*, 39–67. [CrossRef]
41. Toihir, A.M.; Portafaix, T.; Sivakumar, V.; Bencherif, H.; Pazmino, A.; Bègue, N. Variability and trend in ozone over the southern tropics and subtropics. *Ann. Geophys.* **2018**, *36*, 381–404. [CrossRef]
42. NASA Ozone Watch: Latest status of ozone. Available online: https://ozonewatch.gsfc.nasa.gov/ (accessed on 2 February 2020).
43. Harmouche, J.; Fourer, D.; Auger, F.; Borgnat, P.; Flandrin, P. The sliding singular spectrum analysis: A data-driven nonstationary signal decomposition tool. *IEEE Trans. Signal. Process.* **2017**, *66*, 251–263. [CrossRef]
44. Moghtaderi, A.; Borgnat, P.; Flandrin, P. Trend filtering: Empirical mode decompositions versus $\ell 1$ and Hodrick–Prescott. *Adv. Adapt. Data Anal.* **2011**, *3*, 41–61. [CrossRef]
45. Geetikaverma, V.S. Empirical Wavelet Transform & its Comparison with Empirical Mode Decomposition: A review. *Int. J. Appl. Eng.* **2016**, *4*, 5.
46. Wu, Z.; Huang, N.E. A study of the characteristics of white noise using the empirical mode decomposition method. *P. R. Soc. A-Math. Phy.* **2004**, *460*, 1597–1611. [CrossRef]
47. Sang, Y.-F.; Wang, Z.; Liu, C. Comparison of the MK test and EMD method for trend identification in hydrological time series. *J. Hydrol.* **2014**, *510*, 293–298. [CrossRef]
48. Taylor, K.E. Summarizing multiple aspects of model performance in a single diagram. *J. Geophys. Res. Atmos.* **2001**, *106*, 7183–7192. [CrossRef]
49. Bègue, N.; Vignelles, D.; Berthet, G.; Portafaix, T.; Payen, G.; Jégou, F.; Benchérif, H.; Jumelet, J.; Vernier, J.-P.; Lurton, T.; et al. Long-range transport of stratospheric aerosols in the Southern Hemisphere following the 2015 Calbuco eruption. *Atmos. Chem. Phys.* **2017**, *17*, 15019–15036. [CrossRef]
50. Bègue, N.; Shikwambana, L.; Bencherif, H.; Pallotta, J.; Sivakumar, V.; Wolfram, E.; Mbatha, N.; Orte, F.; Du Preez, J.; Ranaivombola, M.; et al. Statistical analysis of the long-range transport of the 2015 Calbuco volcanic eruption from ground-based and space-borne observations. *Ann. Geophys.* **2020**, *38*, 395–420.

Article

Stratosphere–Troposphere Exchange and O_3 Variability in the Lower Stratosphere and Upper Troposphere over the Irene SHADOZ Site, South Africa

Thumeka Mkololo [1,2,*], Nkanyiso Mbatha [3], Venkataraman Sivakumar [1], Nelson Bègue [4], Gerrie Coetzee [2] and Casper Labuschagne [2]

1 School of Chemistry and Physics, University of KwaZulu Natal, Private Bag x 54001, Westville Campus, Durban 4000, South Africa; venkataramans@ukzn.ac.za
2 South African Weather Service, Global Atmosphere Watch Station, P.O. Box 320, Stellenbosch 7599, South Africa; gerrie.coetzee@weathersa.co.za (G.C.); casper.labuschagne@weathersa.co.za (C.L.)
3 Department of Geography and Environmental Studies, University of Zululand, KwaDlangezwa 3886, South Africa; mbathanb@unizulu.ac.za
4 Laboratoire de l'Atmosphère et des Cyclones, LACy UMR 8105, Université de la Réunion, 97400 Réunion Island, French; nelson.begue@univ-reunion.fr
* Correspondence: thumeka.mkololo@weathersa.co.za; Tel.: +27-(0)-21-888-2487

Received: 17 February 2020; Accepted: 7 May 2020; Published: 3 June 2020

Abstract: This study aims to investigate the Stratosphere-Troposphere Exchange (STE) events and ozone changes over Irene (25.5° S, 28.1° E). Twelve years of ozonesondes data (2000–2007, 2012–2015) from Irene station operating in the framework of the Southern Hemisphere Additional Ozonesodes (SHADOZ) was used to study the troposphere (0–16 km) and stratosphere (17–28 km) ozone (O_3) vertical profiles. Ozone profiles were grouped into three categories (2000–2003, 2004–2007 and 2012–2015) and average composites were calculated for each category. Fifteen O_3 enhancement events were identified over the study period. These events were observed in all seasons (one event in summer, four events in autumn, five events in winter and five events in spring); however, they predominantly occur in winter and spring. The STE events presented here are observed to be influenced by the Southern Hemisphere polar vortex. To strengthen the investigation into STE events, advected potential vorticity maps were used, which were assimilated using Modélisation Isentrope du transport Méso–échelle de l'Ozone Stratosphérique par Advection (MIMOSA) model for the 350 K (~12–13 km) isentropic level. These maps indicated transport of high latitude air masses responsible for the reduction of the O_3 mole fractions at the lower stratosphere over Irene which coincides with the enhancement of ozone in the upper troposphere. In general, the stratosphere is dominated by higher Modern Retrospective Analysis for Research Application (MERRA-2) potential vorticity (PV) values compared to the troposphere. However, during the STE events, higher PV values from the stratosphere were observed to intrude the troposphere. Ozone decline was observed from 12 km to 24 km with the highest decline occurring from 14 km to 18 km. An average decrease of 6.0% and 9.1% was calculated from 12 to 24 km in 2004–2007 and 2012–2015 respectively, when compared with 2000–2003 average composite. The observed decline occurred in the upper troposphere and lower stratosphere with winter and spring showing more decline compared with summer and autumn.

Keywords: ozone enhancement; Irene; ozone decline; potential vorticity; ozonesondes

1. Introduction

The stratosphere and troposphere have different characteristics that are useful in identifying air movement from the stratosphere to the troposphere and vice versa. The stratosphere is characterized by high ozone (O_3) and potential vorticity (PV). On the other hand, the troposphere is characterized by low O_3 and PV. Approximately 90% of O_3 is found in the stratosphere and only 10% in the troposphere. Most of the O_3 in the stratosphere is situated within the O_3 layer [1] where it plays a critical role in shielding the environment and protecting human health from dangerous ultraviolet (UV) rays. The O_3 hole was discovered over Antarctica in late 1980's [2,3] and its dynamics with Ozone Depleting Substances (ODS) are well documented [4–6]. The appearance of O_3 hole was a big concern because of the relationship that increased UV can have on various processes in the lower troposphere. In 1998, the Montreal protocol was successfully implemented to phase out the use of ODS. Subsequently, ozone was expected to increase in the stratosphere after ODS were phased out. Since 1985, a decreasing trend of 30 DU in total column O_3 was reported for stations over the southern mid-latitudes [7]. A positive trend was only observed in the upper stratosphere above 10 hPa [8,9]. A continuous O_3 decline (even though statistically non-significant) was reported in the lower stratosphere from 1998 until present for the stations lying between 60° N and 60° S [10].

Tropospheric O_3 mole fractions are controlled by chemical and physical processes, and its precursors originate from both natural and anthropogenic activities. In the troposphere, O_3 is a secondary pollutant formed through a number of reactions containing nitrogen oxides (NO_x), volatile organic compounds (VOCs), methane (CH_4), and CO in the presence of sunlight [11] or through the transportation of O_3–rich air from the stratosphere. This process of O_3 movement from the stratosphere to the troposphere and vice versa is known as stratosphere–troposphere O_3 exchange (STE). After the Los Angeles photochemical smog, studies revealed that O_3 can either be transported from the stratosphere or produced from chemical reactions involving O_3 precursors in the troposphere [12]. Stratosphere intrusion (SI) are expected to transport high O_3 and PV to the troposphere. Hence, high O_3, high PV and dry atmospheric air are used to identify stratosphere intrusion events in the troposphere. The STE plays an important role in the chemical budget of O_3 and water vapour of the upper troposphere and lower stratosphere [13]. Stratospheric Intrusion studies are poorly documented in the Southern Hemisphere; hence, it is challenging to find the threshold that could be used to study these events in the literature. Ozone vertical profiles are useful in identifying and studying STE. A number of studies have been undertaken using Irene ozonesondes data. These studies include: (a) study of O_3 climatology over Irene by Diab et al. [14] using 1990 to 1994 and 1998 to 2004 dataset, (b) satellite validation [15,16] of global transport models, (c) a study on stratospheric profile and water vapour in Southern Hemisphere [17], and (d) the study on the Southern Hemisphere tropopause [18]. In their study, Diab et al. [14] reported O_3 enhancement in Irene above 10 km (upper troposphere) which can be related to STEs during late winter. Similar to this study, they also noted the absence of seasonal consistency in the occurrence of these events at a height above 10 km level. The frequency of occurrence of STEs in the Irene SHADOZ data set has never been studied due to limited data availability and the frequency of the launching of ozonesondes. However, according to Diab et al. [14], the STEs events are dominant in winter and spring in the Irene station. In support to these observations, a study by Poulida et al. [19] also found high O_3 mole fractions in the upper troposphere dominating in winter and spring months.

There are limited studies conducted over Southern Africa on STEs. A recent study on high O_3 events was conducted by Mulumba in Nairobi, Congo Basin and Irene using ozonesondes data [20]. However, the focus was more on Nairobi and Congo Basin and little was done with Irene data. Hence, this study focuses on O_3 data observed from Irene station. It is crucial to investigate O_3 enhancement events in the troposphere and determine if such episodes were related to stratosphere–troposphere exchange. Also, events such as Cut–off lows, Rossby waves, Quasi-Biennial Oscillation (QBO) and El Niño-Southern Oscillation (ENSO) are all factors that could potentially play a critical role in STEs [21,22]. For example, QBO affects the troposphere by direct effect of QBO on the tropical or subtropical troposphere [23]. A downward movement of easterly winds is more dominant and much stronger compared to westerlies. Easterlies are represented as negative on the QBO index while positive signal represents westerlies. Another way that QBO affects the troposphere is through polar vortex [24–26] processes. Most researchers have defined the polar vortex as a region of high PV. The stratosphere polar vortex develops in autumn when there is no solar heating in polar regions, strengthens in winter and breaks down in spring as sunlight returns to polar regions [27]. The breakdown of stratospheric polar vortex especially plays an important role in O_3 distribution in high latitudes [22]. Several studies have investigated the role of the southern polar vortex with respect to the middle atmosphere of the southern hemisphere [28–30]

A comprehensive study on STEs was recently conducted at three Southern Hemisphere stations (Davis (69° S, 78° E), Macquarie Island (55° S, 159° E) and Melbourne (38° S, 145° E)) by looking at the statistical analysis of STEs and their impact on tropospheric O_3 [31]. This study has coupled observed STEs with meteorological conditions such as low pressure fronts, cutoff low pressure system, indeterminate meteorology and smoke plumes. A total number of 45, 47 and 72 events were detected in Davis, Macquarie and Melbourne stations, respectively, from ozonesondes data. The majority of events were related to low pressure fronts with fire plumes contributing the least.

A number of researchers have investigated O_3 trends in the lower troposphere O_3 [32–36] and only a few studies have conducted trends analysis at different altitudes of troposphere and the lower stratosphere [9,37]. A study by Granados–Munoz and Leblanc [38] investigated tropospheric trends at different altitudes over California using a procedure similar to that described by Cooper et al. [33]. In their study, linear fits of medians, 5th percentiles and 95th percentiles were done using least squares method. Statistical significant negative trends were observed in the lower troposphere (4–7 km) in winter for the medians and 5th percentiles. On the other hand, a positive significant trend of 0.3 ppb/year was reported for the upper troposphere (7 to 10 km) for the period of 2000 to 2015. A non-significant trend was reported for layers closer to the tropopause whilst negative trends were observed in the lower stratosphere (17 to 19 km). A recent study by Ball [9] reported a continuous decrease of O_3 in the lower stratosphere in the region between 60° N and 60° S, while ozone recovery was observed in the upper stratosphere.

The aim of this study is to utilise the available Irene ozonesonde vertical profile to investigate the STE events that are known to lead to SI occurrence. Thus, this study identifies O_3 enhancement events exceeding monthly 90th percentile composite in the upper troposphere, and investigate whether such episodes were due to SI events. Another objective of this study is to investigate O_3 changes in the upper troposphere and lower stratosphere using Irene ozonesondes data and linear regression for medians, 5th percentiles and 95th percentiles.

2. Method and Data

2.1. Ozonesondes

Irene station soundings started in 1998 during the Southern African Fire Atmospheric Research Initiative (SAFARI) campaign in the African region. Currently, the station operates within the framework of the Southern Hemisphere Additional Ozonesondes (SHADOZ). The main aim of the project was to determine O_3 mole fractions in the troposphere and stratosphere, and also to have a full coverage of O_3 measurements over Southern Hemisphere. Since then, balloon launching continued in Irene with ozonesondes launched every second Wednesday of each month circumstances allowing. Ozone vertical profiles are obtained using electrochemical concentration cell (ECC). The heart of the instrument is the electrochemical cell that interfaces with a radiosonde that transmits back data signals to the ground station receiver. The method used by the electrochemical cell to detect O_3 was discussed extensively by Sivakumar et al. [39]. A total number of 250 ozonesondes were launched over the study period. Irene ozonesondes data was retrieved from SHADOZ website [40]. It is important to note that there is a data gap of approximately four years (2008 to September 2012) in the Irene data due to budget constraints and technical problems with the ground receiver. The program was resumed in 2012 when these issues were resolved.

2.2. MERRA-2 Potential Vorticity

The stratosphere has static stability and is known to contain higher potential vorticity (PV) compared with the troposphere. During the stratosphere intrusion episode, an air mass rich in O_3 and high PV enters the lower stratosphere and upper troposphere. Hence, PV can be used to identify troposphere air mass having a stratosphere origin. The tropopause is defined using a PV value as 2 PVU [39,41]. Therefore, any higher PV events located in the troposphere are associated with stratosphere origin. Other studies, have used PV values of 1.5 PVU as a threshold to identify stratosphere air [42,43]. In this study, a PV value of 2 PVU was used as a threshold for stratospheric air. This study employs PV data from the Modern Retrospective Analysis for Research Application version 2 (MERRA-2) with a spatial resolution of $0.5 \times 0.625°$. MERRA-2 model is an Earth System reanalysis model by National Aeronautics Space Administration (NASA) Global Modeling and Assimilation Office (GMAO). More details about MERRA-2 can be found in the website: https://gmao.gsfc.nasa.gov/reanalysis/MERRA-2/. Although the units of PV are $Km^2\ kg^{-1}s^{-1}$, PV Units (PVU) (where 1 PVU = $1 \times 10^{-6}\ Km^2\ kg^{-1}s^{-1}$) will be used for convenience in this study. This study used NASA instrument Panoply software (see https://www.giss.nasa.gov) to view the vertical slices of PV over Irene [44].

2.3. Data Processing

The ozonesonde data is recorded every two seconds from 1.5 km to approximately 28 km. Ozone averages were calculated from 2 seconds data for each kilometer (km) ascended (e.g., 1 km, 2 km, 3 km to 28 km). As the ozonesonde ascends, pressure, temperature, humidity and O_3 (both ppm and DU units) are also recorded. The current study used the available ozonesondes data to investigate high O_3 events. The ozonesonde data was grouped into months in order to calculate averages, standard deviations, medians, 5th and 95th percentiles from 1 to 28 km. Monthly averaged data was used in conjunction with individual high O_3 event profiles to determine how individual profiles differ from their respective monthly composite profiles. This study defines high O_3 event as the event where O_3 exceeds the monthly 90th and 95th percentile composites. Events exceeding these percentile composites were selected as an observed high O_3 events. However, these events might originate from different sources such as plumes, stratospheric-troposphere exchange (STE) and other man made activities that generate O_3 precursors. Due to this reason, potential vorticity (PV) of 2PV was added as another criteria to identify events of stratospheric origin. PV was used in this study because it is one of the characteristics to differentiate between the stratosphere and troposphere air masses. Monthly 90th and 95th percentile composites were calculated from the available data of all ozonesondes launched during the study period. Any profile exceeding the monthly 90th and 95th percentile composites at a height between 6 and 11 km was considered for further investigation. The main reason to focus between 6 km and 11 km is to eliminate tropospheric pollution and to select events that occurred below the tropopause. Our method differs from previous studies that identified Stratospheric Intrusion (SI) events as an event where O_3 exceeds 80 ppb and then within 3 km above decreases by 20 ppb or more to a value less than 120 ppb [45]. This method was not used because it will miss some of the events observed over Irene due to lower O_3 mole fractions in the Southern Hemisphere. Another study used the 99th percentile as a threshold to study SI O_3 events and their impact on tropospheric ozone [31]. Similar percentile threshold was not applied in this study for similar reason stated above. We attempted to use the 95th percentile as a threshold, however, we missed five events. Consequently, we opted to use 90th percentile as a threshold.

Average O_3 composite profiles were calculated for three categories (2000–2003, 2004–2007 and 2012–2015) using the available ozonesondes data. In the case of annual and seasonal O_3 changes, monthly averaged composites were calculated from the available ozonesondes data. Monthly averaged composites were used to compensate data gaps that occurred in summer and autumn. Hence, only summer and autumn composites were used to fill 2015 data gaps. This exercise was done to prevent the bias that may be caused by months with data gaps in calculating O_3 changes at different altitudes. Data gaps were covered only for 2015, not for the years where there was no ozonesondes data for the complete year. Annual changes were studied by averaging monthly data into yearly averages at different layers such as 13–15 km, 16–18 km, 19–21 km and 22–24 km. Furthermore, the medians, 5th and 95th percentiles were calculated at each layer. The slope was determined by fitting a linear trend to yearly averaged data plotted on the scatter plot. The standard error corresponding to the slope was calculated at each layer for median values, and both the 5th and 95th percentile. Similar approach was used to calculate seasonal O_3 changes.

3. Results and Discussion

Figure 1 indicates the monthly (a) and yearly (b) ozonesondes data launched from year 2000 to year 2015 at Irene station. Over the study period, a total number of 250 ozonesondes were launched from January 2008 to September 2012 showing a significant data gap. However, 12.4% of the launched ozonesondes did not reach above 28 km. A maximum of 25 ozonesondes were launched in October and November, respectively whilst a total of 12 ozonesondes were launched in January over the study period. On a per annual basis, the highest number (39) of ozonesondes were launched in 2000 while 2015 reflects the lowest number (11) of ozonesondes launched. The discrepancy in the annual number of ozonesondes launched was a factor of budgetary constraints as well as some operational issues encountered. In general, the target for this SHADOZ station is to launch at least two ozonesondes per month, which makes a total of 24 launches per year.

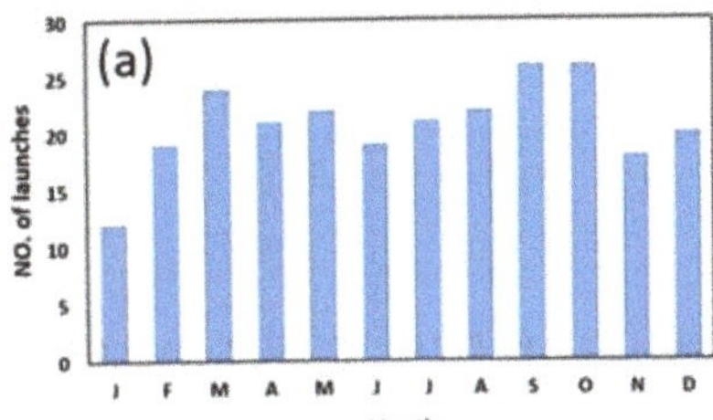

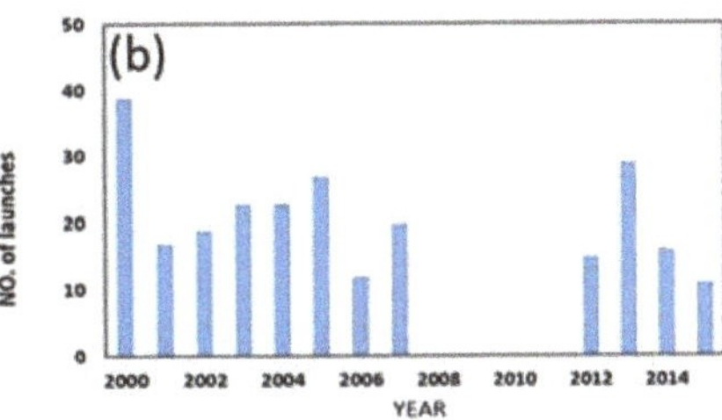

Figure 1. Number of ozonesondes launched at Irene from 2000 to 2015, expressed (**a**) monthly and (**b**) annually.

Figure 2. shows O_3 vertical profiles for the troposphere (Figure 2a) and stratosphere (Figure 2b). The data was grouped into three categories namely 2000–2003, 2004–2007, 2012–2015, and the O_3 composite was calculated at kilometre intervals from 1 km to 28 km. The data used in Figure 2 also include 12.4% of ozonesondes that didn't reach 28 km. Therefore, 87.6% of ozonesondes launched over the study period reached 28 km. Three vertical profiles (2000–2003, 2004–2007 and 2012–2015) were compared and the difference between 2000–2003 and 2004–2007 and between 2000–2003 and 2012–2015 was calculated to determine O_3 variation over the years. The percentage decrease for 2004–2007 profiles were calculated from O_3 difference between 2000–2003 and 2004–2007, similarly, the percentage decrease for 2012–2015 was calculated from 2000–2004 and 2012–2015 O_3 difference. A continuous decrease from 2004–2007 and 2012–2015 was apparent at the height between 12 km and 28 km, with highest decrease occurring at a height between 14 km and 18 km. An average decrease of 6.0% and 9.1% was observed at a height between 12 km and 26 km for 2004–2007 and 2012–2015, respectively, when compared to the 2000–2003 period. These results are in agreement with a study by Ball et al. [10] which reported evidence from multiple satellite measurements that ozone in the lower stratosphere between 60° S and 60° N has indeed continued to decline since 1998.

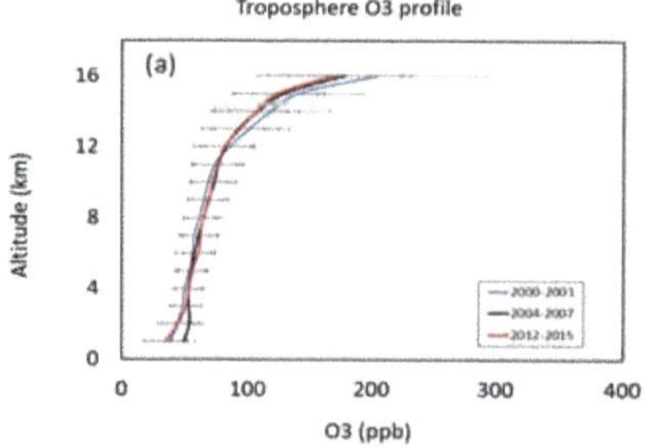

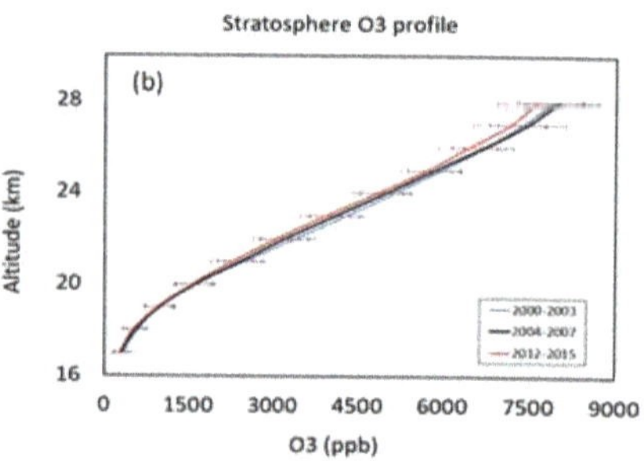

Figure 2. Ozone troposphere (**a**) and stratosphere (**b**) profiles. The blue line indicates 2000–2003 averages, black line indicates 2004–2007 averages and red line indicates 2012–2015 averages. The error bars indicate the standard deviation.

Figure 3 shows the O_3 seasonal vertical profiles averages for the troposphere and stratosphere. In the year 2012–2015, summer experienced high O_3 increase from 5 km to 12 km (Figure 3a). On the other hand, 2000–2007 experienced high O_3 mole fractions in the lower troposphere (Figure 3a,c,e,g), with the increase noted up to 8 km during the winter (Figure 3e). On the other hand, O_3 decrease was observed in the upper troposphere and lower stratosphere, with 2012–2015 experiencing lower mole fractions compared to 2004–2007 and 2000–2003 (Figure 3b,d,f,h). In general, a continuous non-consistent O_3 decrease occurred from the upper troposphere (above 16 km) to lower stratosphere (28 km) for all seasons. These results suggest that the observed O_3 decrease in the lower stratosphere is independent of season. Whereas, O_3 increase between 1.5 and 4 km could be related to increase in urban influence boundary layer precursors [46]. Such O_3 precursors could originate from domestic heating and power stations. Maximum standard deviation is observed at altitudes closer to tropopause region. Such increase could be related to STE and other dynamic changes occurring in the tropopause region. This variation was observed to be lesser in summer when compared to other seasons. Sivakumar and Ogunniyi [38] reported similar observations of higher standard deviation closer to the tropopause height.

Tables 1 and 2 summarizes statistic of high O_3 events that were obtained by using the 95th and 90th percentile composites and potential vorticity of 2 PVU as thresholds. All episodes that exceed the monthly 95th percentile composite and 2 PVU were automatically classified as high ozone events (Table 1). However, it was noted that more episodes could be identified when 90th percentile composite is used instead of 95th percentile (Table 2). The monthly average composites, 90th percentile and 95th percentile composites were calculated by using 2000–2015 Irene ozonesondes data. The maximum of the peak was defined as the highest ozone observed from a particular ozone profile at a particular altitude. In this case, it is the maximum of the profile dated on the first column. Delta ozone was defined as the difference between the maximum of a particular event observed at a particular altitude and 90th or 95th percentile composite profile. The altitude of the event was defined as an altitude where maximum values of O_3 occurred. Based on observations indicated in Table 2, it is noted that stratospheric intrusions can reach 7 km altitude over Irene. Similar results of the occurrence of deep intrusions at 7 km altitude were reported at Reunion Island [47]. Clain et al. [47] used different PV thresholds (1.0, 1.5 and 2.0 PVU) on the study of STEs in Reunion Island. Their findings revealed that the number of detected STEs depends on the PV value and duration of back trajectories. About 9.9% STEs were detected using 2 PVU and 2 days back trajectories relative to 28.5% STEs detected using 1 PVU and 10 days back trajectories. In this study, 6.8% STEs were detected using 2 PVU and 90th percentile composite as a threshold. While 4.6% STEs were detected using 2 PVU and 95th percentile composite as a threshold. Figures 4–10b show more events of high PV propagation from the stratosphere to the troposphere. However, in most cases there were some data gaps of ozonesondes.

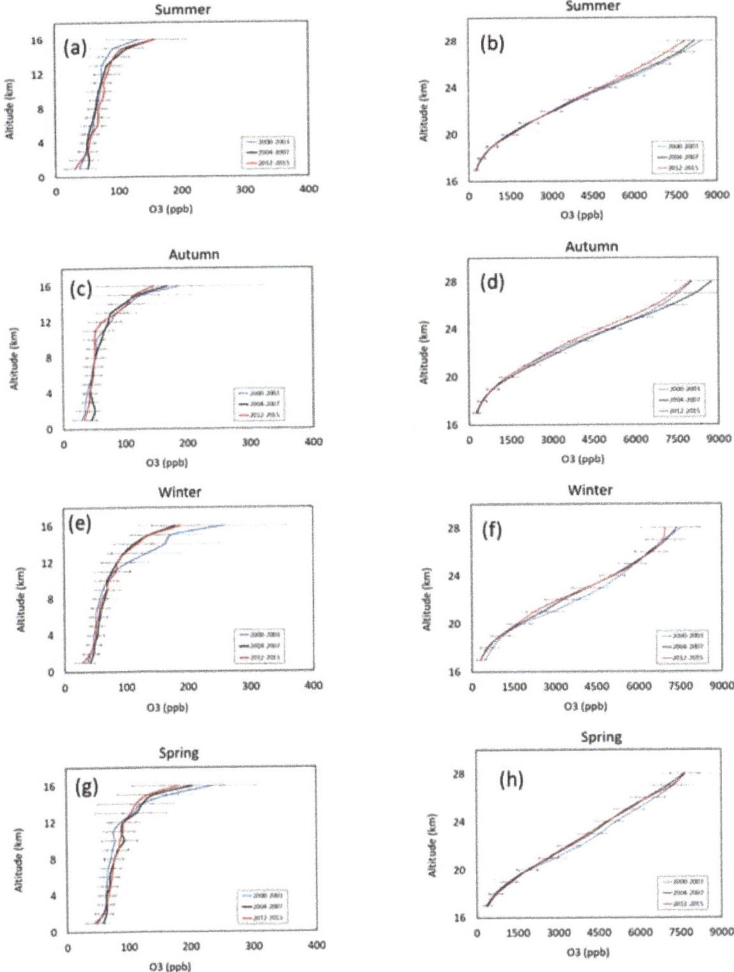

Figure 3. Ozone seasonal profiles for the troposphere (**a**,**c**,**e**,**g**) and stratosphere (**b**,**d**,**f**,**h**). Blue line indicates 2000–2003 averages, black line indicates 2004–2007 averages and red line indicates 2012–2015 averages. The error bars indicate the standard deviation.

Table 1. Summary of high ozone (ppb) events statistics using 95th percentile and potential vorticity (2 PVU) as thresholds.

Date of the Event	Monthly Composite	Monthly 95th Percentile Composite	Maximum Peak of the Event	Delta Ozone	Altitude of the Event
07/08/2002	70.00	99.00	160.50	61.50	11
07/07/2004	55.68, 67.62	74.00, 147.00	146.73, 215.51	72.73, 68.51	9, 11
15/09/2004	79.15	103.00	144.02	41.02	9
26/08/2005	71.98	99.00	171.37	72.37	11
12/04/2006	58.03	110.80	120.00	9.20	11
08/08/2007	58.18	99.00	105.71	6.71	9
10/01/2007	74.65	93.00	110.91	17.91	11
31/07/2013	60.41	90.50	121.67	31.17	10
16/04/2014	50.42	72.00	81.46	9.46	9
25/11/2015	91.67	134.00	162.03	28.03	10

Table 2. Summary of high ozone (ppb) events statistics using 90th percentile and potential vorticity (2 PVU) as thresholds.

Date of the Event	Monthly Composite	Monthly 90th Percentile Composite	Maximum Peak of the Event	Delta Ozone	Altitude of the Event
07/08/2002	70.00	96.00	160.50	64.50	11
27/11/2002	87.33	99.00	117.13	18.13	9
03/03/2004	64.91	82.00	87.13	5.13	9
07/07/2004	55.68, 67.62	70.00, 86.00	146.73, 215.51	76.73, 129.51	9, 11
15/09/2004	79.15	99.00	144.02	45.02	9
26/08/2005	71.98	96.00	171.37	75.37	11
12/04/2006	58.03	71.00	120.00	49.00	11
01/03/2007	58.45	74.00	81.27	7.27	7
08/08/2007	58.18	76.00	105.71	29.71	9
10/01/2007	74.65	87.00	110.91	23.91	11
03/10/2012	89.50	112.00	110.94	2.94	10
17/10/2012	89.50	112.00	137.00	25.00	10
31/07/2013	60.41	72.00	121.67	49.67	10
16/04/2014	50.42	66.00	81.46	15.46	9
25/11/2015	91.67	113.00	162.03	49.03	10

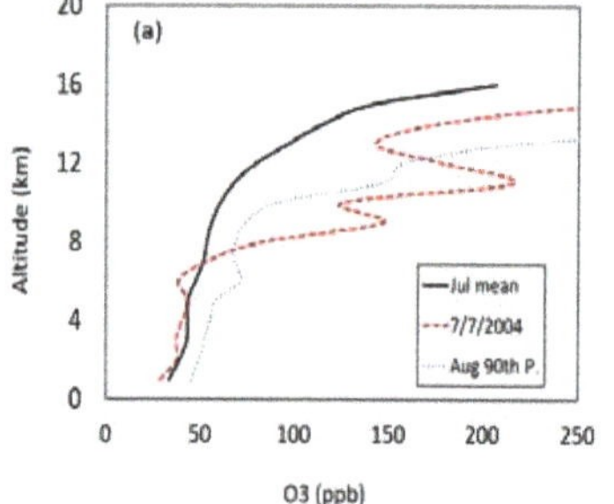

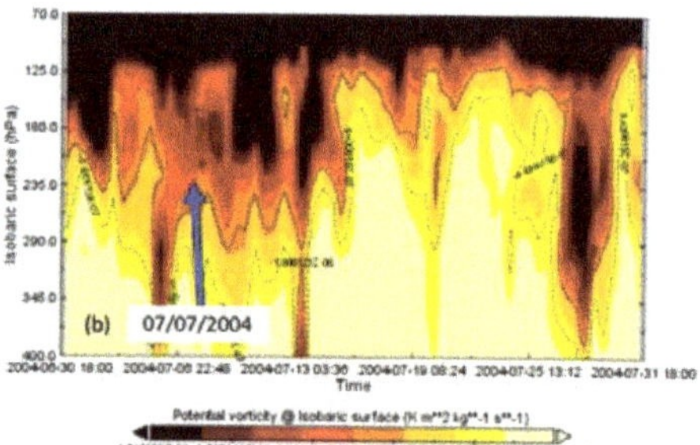

Figure 4. (**a**) Ozone troposphere profile over Irene on 07 July 2004. (**b**) Potential vorticity at Isobaric surface over Irene on July 2004 (from MERRA-2). Solid line in Figure 4a indicates monthly O_3 average composite, broken line indicates O_3 event and dotted line indicate 90th percentile composite. Colours in Figure 4b indicates the level of potential vorticity, black indicates high potential vorticity of more than 3 Km^{-2} $kg^{-1}s^{-1}$ while yellow indicates potential vorticity of less than 2 Km^{-2} $kg^{-1}s^{-1}$. Blue arrow in Figure 4b indicates the event that is associated with O_3 profile in Figure 4a.

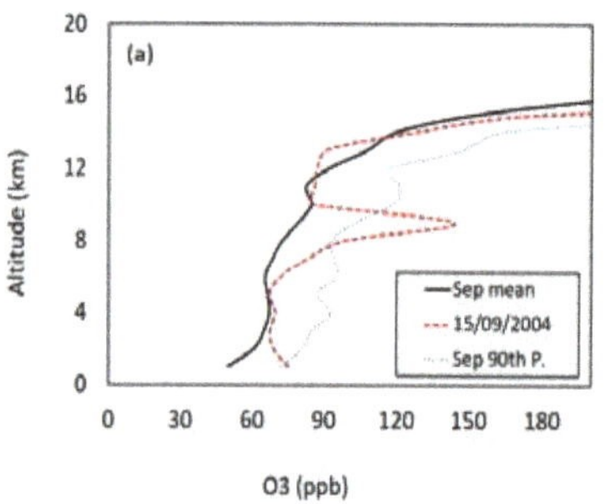

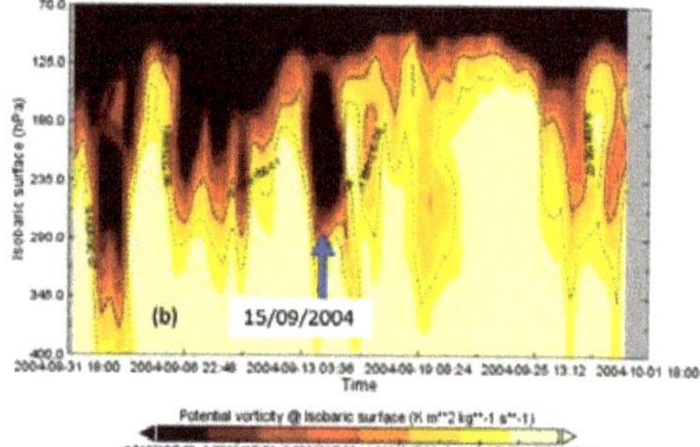

Figure 5. (**a**) Ozone troposphere profile over Irene on 15 September 2004. (**b**) Potential vorticity at Isobaric surface over Irene on July 2004 (from MERRA-2). Solid line in Figure 5a indicates monthly O_3 average composite, broken line indicates O_3 event and dotted line indicate 90th percentile composite. Colours in Figure 5b indicates the level of potential vorticity, black indicates high potential vorticity of more than 3 Km^{-2} $kg^{-1}s^{-1}$ while yellow indicates potential vorticity of less than 2 Km^{-2} $kg^{-1}s^{-1}$. Blue arrow in Figure 5b indicates the event that is associated with O_3 profile in Figure 5a.

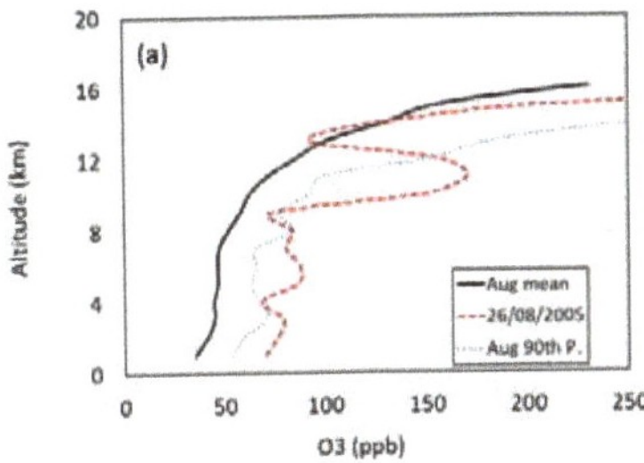

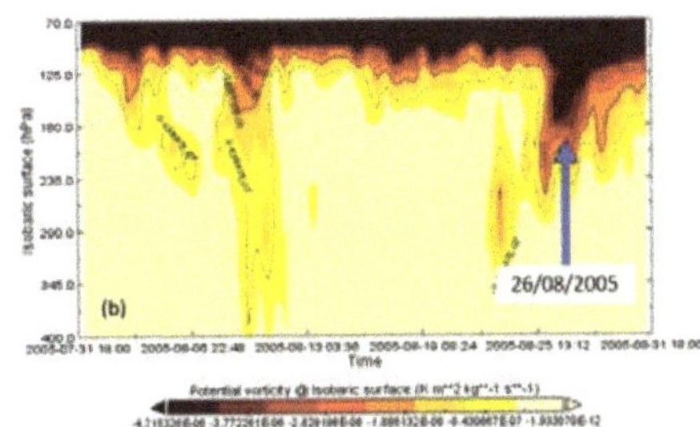

Figure 6. (**a**) Ozone troposphere profile over Irene on 26 August 2005. (**b**) Potential vorticity at Isobaric surface over Irene on July 2004 (from MERRA-2). Solid line in Figure 6a indicates monthly O_3 average composite, broken line indicates O_3 event and dotted line indicate 90th percentile composite. Colours in Figure 6b indicates the level of potential vorticity, black indicates high potential vorticity of more than 3 Km^{-2} $kg^{-1}s^{-1}$ while yellow indicates potential vorticity of less than 2 Km^{-2} $kg^{-1}s^{-1}$. Blue arrow in Figure 6b indicates the event that is associated with O_3 profile in Figure 6a.

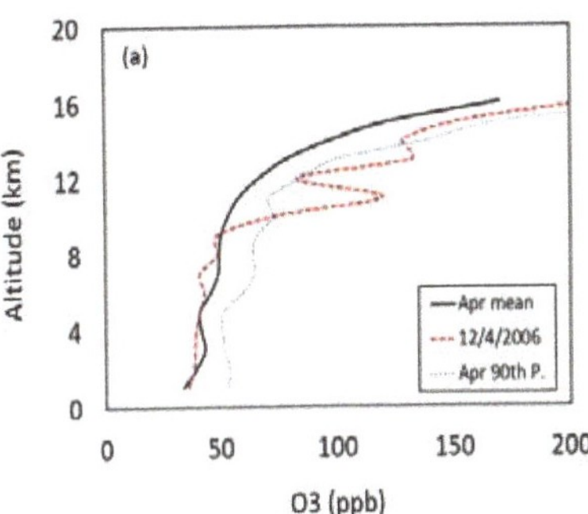

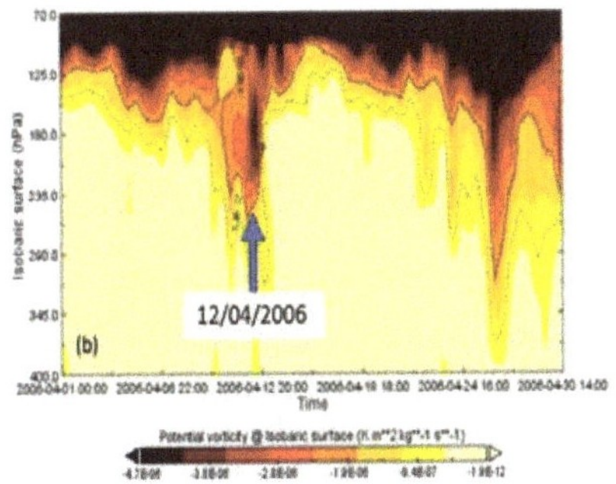

Figure 7. (**a**) Ozone troposphere profile over Irene on 12 April 2006. (**b**) Potential vorticity at Isobaric surface over Irene on July 2004 (from MERRA-2). Solid line in Figure 7a indicates monthly O_3 average composite, broken line indicates O_3 event and dotted line indicate 90th percentile composite. Colours in Figure 7b indicates the level of potential vorticity, black indicates high potential vorticity of more than 3 Km^{-2} $kg^{-1}s^{-1}$ while yellow indicates potential vorticity of less than 2 Km^{-2} $kg^{-1}s^{-1}$. Blue arrow in Figure 7b indicates the event that is associated with O_3 profile in Figure 7a.

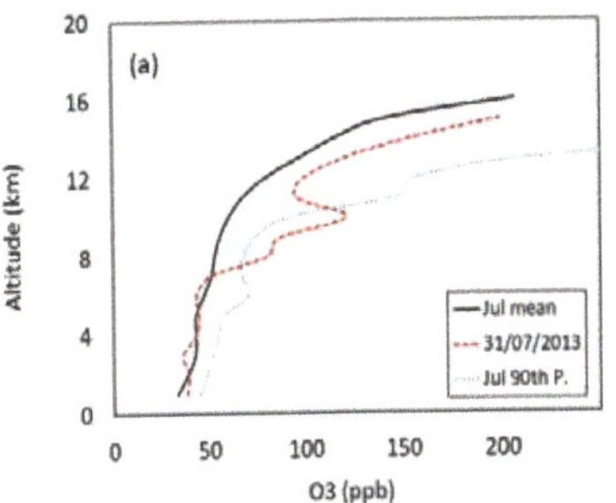

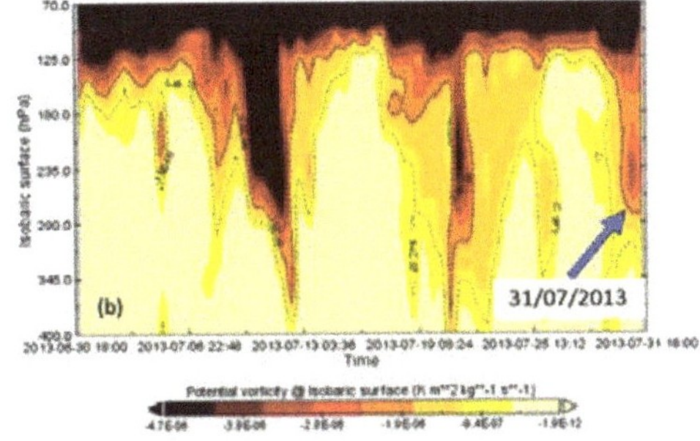

Figure 8. (**a**) Ozone troposphere profile over Irene on 31 July 2013. (**b**) Potential vorticity at Isobaric surface over Irene on July 2004 (from MERRA-2). Solid line in Figure 8a indicates monthly O_3 average composite, broken line indicates O_3 event and dotted line indicate 90th percentile composite. Colours in Figure 8b indicates the level of potential vorticity, black indicates high potential vorticity of more than 3 Km^{-2} $kg^{-1}s^{-1}$ while yellow indicates potential vorticity of less than 2 Km^{-2} $kg^{-1}s^{-1}$. Blue arrow in Figure 8b indicates the event that is associated with O_3 profile in Figure 8a.

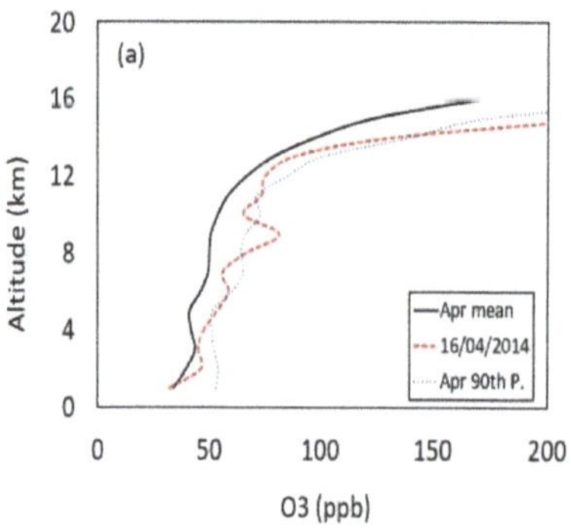

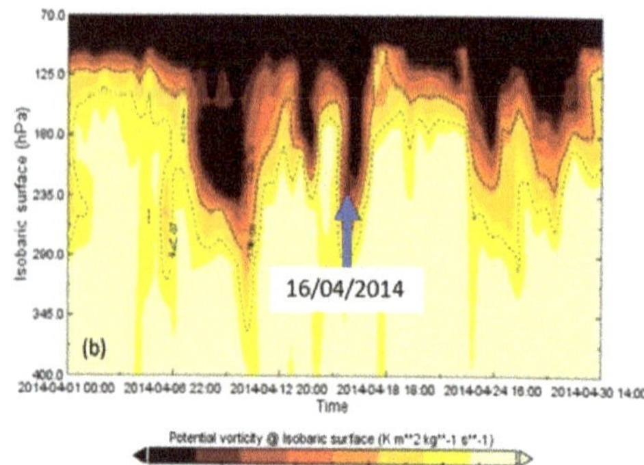

Figure 9. (**a**) Ozone troposphere profile over Irene on 16 April 2014. (**b**) Potential vorticity at Isobaric surface over Irene on July 2004 (from MERRA-2). Solid line in Figure 9a indicates monthly O_3 average composite, broken line indicates O_3 event and dotted line indicate 90th percentile composite. Colours in Figure 9b indicates the level of potential vorticity, black indicates high potential vorticity of more than 3 Km^{-2} $kg^{-1}s^{-1}$ while yellow indicates potential vorticity of less than 2 Km^{-2} $kg^{-1}s^{-1}$. Blue arrow in Figure 9b indicates the event that is associated with O_3 profile in Figure 9a.

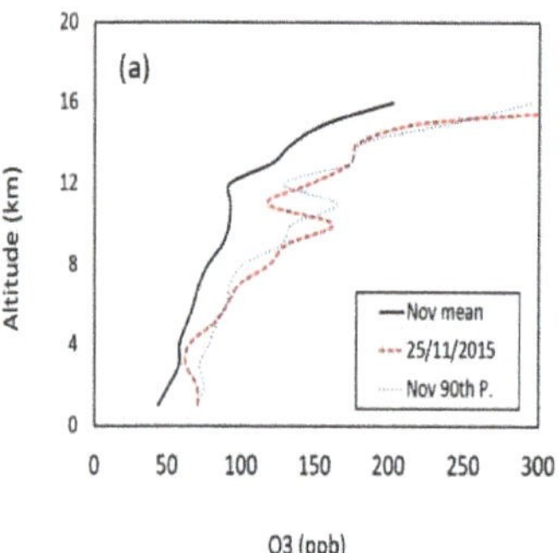

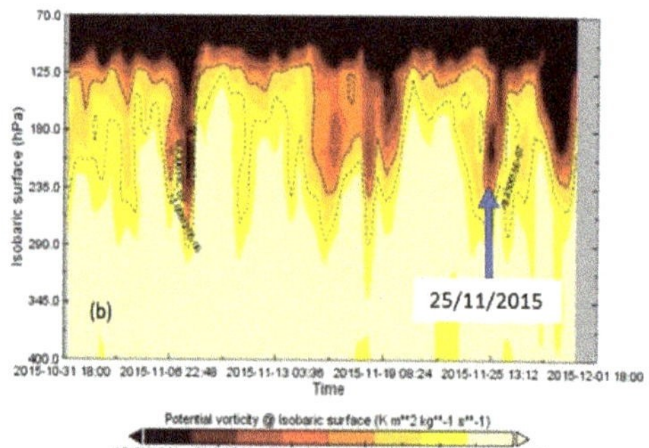

Figure 10. (**a**) Ozone troposphere profile over Irene on 25 November 2015. (**b**) Potential vorticity at Isobaric surface over Irene on July 2004 (from MERRA-2). Solid line in Figure 10a indicates monthly O_3 average composite, broken line indicates O_3 event and dotted line indicate 90th percentile composite. Colours in Figure 10b indicates the level of potential vorticity, black indicates high potential vorticity of more than 3 Km^{-2} $kg^{-1}s^{-1}$while yellow indicates potential vorticity of less than 2 Km^{-2} $kg^{-1}s^{-1}$. Blue arrow in Figure 10b indicates the event that is associated with O_3 profile in Figure 10a.

3.1. *High O_3 Events*

In this study, events with high O_3 mole fractions exceeding monthly 90th percentile composites were selected and discussed in terms of PV retrieved using MERRA-2 reanalysis system. As indicated on a PV chart (PV plotted on an isotropic surface) in Figure 4, values outside the contour lines appear relatively low with PV values of about -1.9×10^{-12} Km^{-2} $kg^{-1}s^{-1}$ compared with values inside the contour lines, which have PV values ranging between -4.7×10^{-6} to 1.9×10^{-6} Km^{-2} $kg^{-1}s^{-1}$. Higher PV values in the troposphere indicate air mass of stratospheric origin due to increased static stability. PV is generally negative in the Southern Hemisphere (SH) and is usually multiplied by negative one (−1) to appear positive [48,49]. In general, higher values of PV are found in the stratosphere than in the troposphere.

Stratosphere polar vortex forms during autumn in Southern Hemisphere [50]. Moreover, with regards to the geographic position of Irene, this location experiences anticyclonic gyre due to midlatitude westerly waves that occur in autumn and winter [51]. These anticyclonic gyres are responsible for increase in pollutant concentrations for a long period [14]. Two possible stratosphere–troposphere O_3 autumn events (12 April 2006 and 16 April 2014) were selected for discussion.

For the purpose of this study, three possible stratosphere–troposphere O_3 winter events (7 July 2004, 26 August 2005 and 31 July 2013) were selected for discussion. Generally, one may conclude that activities such as anticyclone patterns, domestic usage of biofuels for heating, power generating plants and STEs are the cause of higher O_3 enhancement during this period in Irene [51]. Furthermore, during this season, stratospheric polar vortex starts to be very active in polar region [50].

Also, in this study, two high O_3 events (15 September 2004 and 25 November 2015) were selected in spring and discussed. It is well known that the spring O_3 enhancement events can either be caused by biogenic emissions, biomass burning and lightning production or a combination of them all [11]. However, the occurrence of STEs is dominant in winter and spring over the study area [14], and during this time, the stratospheric polar vortex is more active. According to Clain et al. [47], it is possible that O_3 mole fractions related to stratosphere intrusion can be influenced by climatological O_3 background.

3.2. *Case Studies on High Ozone Events*

Figures 4a–10a show events that were selected for the case study. These events are part of the fifteen high O_3 events that were identified over the study period. Since these events took place in the month of April, July, August and September, 90th percentile composites for these months were used as thresholds. On these days, high O_3 peaks exceeding the monthly 90th percentile composites were observed between 9 km and 11 km. Figures 4b–10b show MERRA-2 potential vorticity plotted against the Isobaric surface between 70 hPa and 400 hPa. These vertical PV slices are for the whole month for the selected events, and averaged to the closest latitude and longitude to Irene SHADOZ site. As indicated on the PV slices, there are several episodes observed in these months where air masses with higher PV of approximately 3.0 PVU propagated from the lower stratosphere (70 hPa) to the upper troposphere (400 hPa). These events are shown as the downward tongues on the PV slices. However, for the purpose of this study, we focus on the time scale closer to the event dates. Blue arrows in PV slices indicate the events that are associated with high O_3 in vertical profiles. As indicated by O_3 vertical profiles, the observed high O_3 events coincides with high PV observed in the higher troposphere. Therefore, it can be reasonably concluded that the observed high O_3 in the upper troposphere could be of stratospheric origin.

3.3. *Dynamical Context Using MIMOSA Model*

There are several studies that have shown that the dynamics of the Southern Hemisphere polar vortex has an influence in the nearby surrounding structures (upper troposphere and stratosphere) of the Southern Hemisphere [28–30]. A useful method that can assist in profiling the isentropic transport across the dynamical barriers in the stratosphere is the MIMOSA (Modélisation Isentrope du transport Méso–échelle de l'Ozone Stratosphérique par Advection) model. MIMOSA model is a high-resolution advection contour model that is based on Ertel's potential vorticity which was developed at the Service d'Aeronomie by Hauchecorne et al. [52]. The advection is driven by ECMWF meteorological analyses at a resolution of $0.5° \times 0.5°$. In the case of the PV, its slow adiabatic evolution is taken into account by relaxing the model PV towards the PV calculated from the ECMWF fields with a relaxation time of 10 days. Using this procedure, it is possible to run the model continuously and follow the evolution of PV filaments during several months. This model system enables the investigation of the contribution of the horizontal transport mechanism in the vertical distribution of ozone over high latitudes, mid-latitudes and subtropics. The model gives as an output the advected potential vorticity (APV) with a resolution of $0.3° \times 0.3°$ which is measured in potential vorticity units (PVU) which corresponds to 1×10^{-6} Km^{-2} $kg^{-1}s^{-1}$.

In their recent study, Orte et al. [30] successfully showed the influence of the polar vortex over Rio Gallegos, Argentina by using the APV calculated from the MIMOSA high-resolution advection model. Having adopted a similar approach in this study, the influence of the dynamics of the polar vortex over Irene during the days where the STE was observed is also investigated using the APV outputs from the MIMOSA model. The APV maps assimilated using MIMOSA model for the 350 K isentropic level plotted for the 07 July 2004 (a), 15 September 2004 (b), 26 August 2005 (c), 12 April 2006 (d), 30 July 2013 (e), 31 July 2013, 16 April 2014 (f), and 25 November 2015 (g) are shown in Figure 11. The location of Irene is indicated by a black dot in the maps. The slices of APV for 350 K isentropic level which is equivalent to 12–13 km were selected because this is found to be an appropriate pressure level to investigate an STE event.

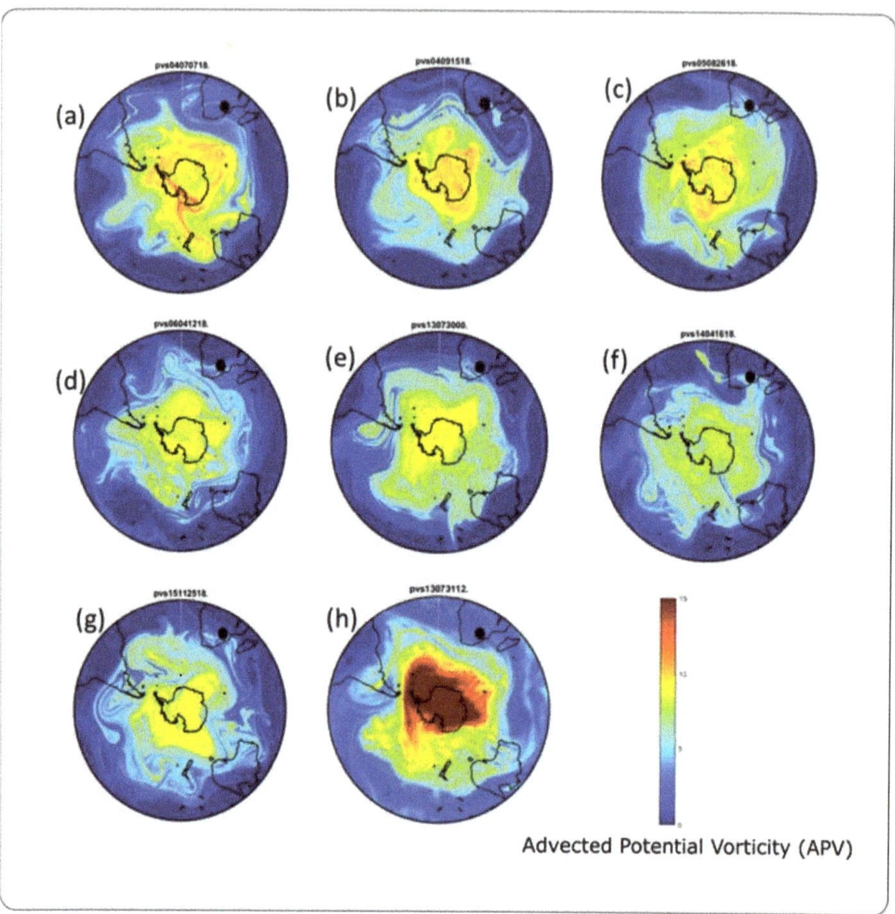

Figure 11. Advected Potential Vorticity (APV) maps assimilated with the MIMOSA model for the 350 K isentropic level (in PVU) and for 07 July 2004 (**a**), 15 September 2004 (**b**), 26 August 2005 (**c**), 12 April 2006 (**d**), 30 July 2013 (**e**), 31 July 2013 (**f**), 16 April 2014 (**g**), and 25 November 2015 (**h**). The black dot represents the location of the Irene site.

In general, during all these days which experienced STE process there is obvious passing of APV values with an averaged value of 8 PVU over Irene, South Africa. This is confirmed by blue tongue at the 350 K isentropic level that reflect higher PV values passing over Irene during the days of the STE that were profiled in this study. It can thus be reasoned that this isentropic transport seems to be responsible for the observed reduction of the O_3 mole fractions at the lower stratosphere over Irene which takes place at the same time with the enhancement of ozone in the upper troposphere. The possible dynamical event which is well simulated by the MIMOSA model during the STEs presented here could be that the high APV values are transported from the high latitudes towards the tropics bringing air masses that contain lower ozone concentrations. There is an upward propagation of the middle atmosphere planetary waves in the high and mid-latitudes regions which results to downward propagation around the lower latitudes of the lower stratosphere [28]. The O_3 mole fractions in the lower stratosphere are transported to the upper troposphere, and hence the observed STEs. The influence of the high latitude stratospheric air masses over Irene were also reported on by Semane et al. [28] via their study of the dynamics of the middle atmosphere during the winter of year 2002. Besides, this was a special winter in the southern hemisphere because of the unprecedented year 2002 major stratospheric warming [53,54].

Diab et al. [14] reported a tropopause folding event that occurs during the winter and spring transition period in Irene. The tropopause folding is a good indicator of a STE physical process [14]. Thus, with an improvement of SHADOZ data collection at Irene site since then, it is always important to investigate such a physical process in this study. Figure 12 shows monthly averaged composite of O_3 vertical profiles measured at Irene for the year period from 2000 to 2015. These profiles were plotted for the height region ranging from 1 km to 15 km for January (Jan) to December (Dec). There is a general significant intrusion of higher O_3 mole fractions which are sourced from the stratosphere which is observed in late winter and spring months. In their study, Diab et al. associated this O_3 injection to middle troposphere with westerly winds, which marks the end of maritime season [14]. Also, the subtropical jet was reported to play a role in permitting ozone-rich stratospheric air to penetrate into the troposphere [55]. While most of the free troposphere over Irene was characterised by O_3 mole fraction of approximately 55–60 ppb, the late winter months experience an increase of O_3 concentration to approximately 80 ppb just above the planetary boundary layer. It is also worth noting that the spring season is the period where there are activities such as anthropogenic pollutants sourced from the Congo region and Highveld region biomass burning, and natural activities such as lightning from rainy season and biogenic activities [56]. On the other hand, a similar observation to that which was reported by Diab et al. [14], the tropical tropopause layer (TTL) with O_3 mole fraction ranging between 95 ppb and 100 ppb at a height above 14 km from January to February, while it noticeably declined throughout the year, and approached its minimum altitude of 11 km in October.

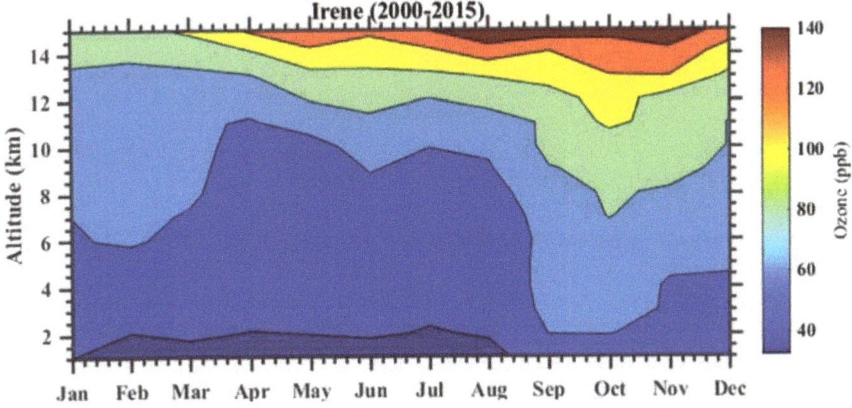

Figure 12. Contour plot of Irene O_3 mole fraction (in ppb) for the period, 2000 to 2015.

3.4. Ozone Decline in Lower Stratosphere

The recovery of O_3 in the upper stratosphere has been well discussed [9,57–61]. However, the investigation of O_3 recovery at different altitude starting from the lower stratosphere upwards still needs attention. This also arise because some recent studies seem to have reported that there may be a continuous decline of O_3 at the lower stratosphere [10,62]. A recent study by Sivakumar and Ogunniyi [38] divided ozonesondes data into two categories, namely tropospheric (0–15 km) and stratospheric region (15–30 km) and reported O_3 maximum occurrence between 22–27 km. Thus, in this study, we also investigate the O_3 decline in the lower stratosphere by using Irene ozonesondes data. Ozone decline is calculated by using medians, 5th and 95th percentiles. Moreover, a composite was calculated at different altitudes (e.g., 13–15 km, 16–18 km and 19–21 km) of the upper troposphere and stratosphere. Rate of change was then calculated by fitting a linear trend on the graphs.

3.4.1. Annual Changes at Different Altitudes

Table 3 summarises the statistics calculated for the medians, 5th and 95th percentiles. Layers corresponding to the upper troposphere (7–9 km) show a positive change of 0.33 ± 0.57, 0.19 ± 0.56 and 0.38 ± 0.88ppb/year for the median, 5th and 95th percentiles respectively. Similarly, at 10–12 km height, a positive change of 0.29 ± 0.58 and 0.24 ± 1.25 ppb/year was observed for the median and 95th percentile respectively. In contrast, a negative change of 2.58 ± 3.90, −0.59 ± 3.17 and −9.63 ± 9.27 ppb/year was observed at 16–18 km for the median, 5th and 95th percentile respectively. Similarly, negative changes were observed at 19–21 and 22–24 km for median, 5th and 95th percentiles. In summary, the results presented here indicate that there are negative changes in the lower stratosphere, while the upper troposphere shows a positive change. Similar observations of O_3 decline in the lower stratosphere were reported by Granados–Munoz and Leblanc [37] when studying tropospheric O_3 seasonal and long-term variability at the JPL–Table Mountain. Furthermore, Ball et al. [10] suggested that lower stratosphere decline contributes to the observed total column O_3 decline. Therefore, the results presented here are consistent with the previous observations reported in literature [10,37].

Table 3. Statistical analysis of ozone at different altitudes.

Altitude	Median [ppb/year]	5th Percentile [ppb/year]	95th Percentile [ppb/year]
7–9 km	0.33 ± 0.57	0.19 ± 0.56	0.38 ± 0.88
10–12 km	0.29 ± 0.58	−0.08 ± 0.78	0.24 ± 1.25
13–15 km	0.21 ± 1.04	0.47 ± 1.13	−2.38 ± 3.28
16–18 km	−2.58 ± 3.90	−0.59 ± 3.17	−9.63 ± 9.27
19–21 km	−6.95 ± 13.07	−7.04 ± 9.83	−9.46 ± 18.31
22–24 km	−16.16 ± 21.83	−21.19 ± 27.39	−14.81 ± 21.82

3.4.2. Seasonal Changes at Different Altitudes

Figure 13 indicates O_3 changes calculated using linear regression at different altitudes during summer, autumn, winter and spring season. And, Table 4(a,b,c,d) provide summery statistics (Median, 5th and 95th percentiles) of these changes. Standard deviation values close to zero indicate O_3 observed over the years was close to the calculated mean. The observations in these tables can be summarised as follows:

7–9 km layer: there was a negative change identified in autumn (−0.11 ± 0.53 ppb/year). While there was a positive change observed in summer (0.53 ± 0.40 ppb/year), winter (0.85 ± 0.72 ppb/year) and spring (0.04 ± 0.57 ppb/year) for the medians. Similarly, there was a positive change observed for the 5th percentiles in summer (0.60 ± 0.58 ppb/year), autumn (0.40 ± 0.34 ppb/year) and winter (0.02 ± 0.17 ppb/year). While spring (−0.02 ± 0.22 ppb/year) showed a negative change. There was a negative change observed in spring (−0.54 ± 1.00 ppb/year), while a positive change was observed in summer (0.23 ± 0.78 ppb/year), autumn (0.68 ± 0.79 ppb/year) and winter (1.14 ± 0.94 ppb/year) for the

95th percentiles. Therefore, it can be concluded that an overall positive change is dominant in this layer for most of the seasons.

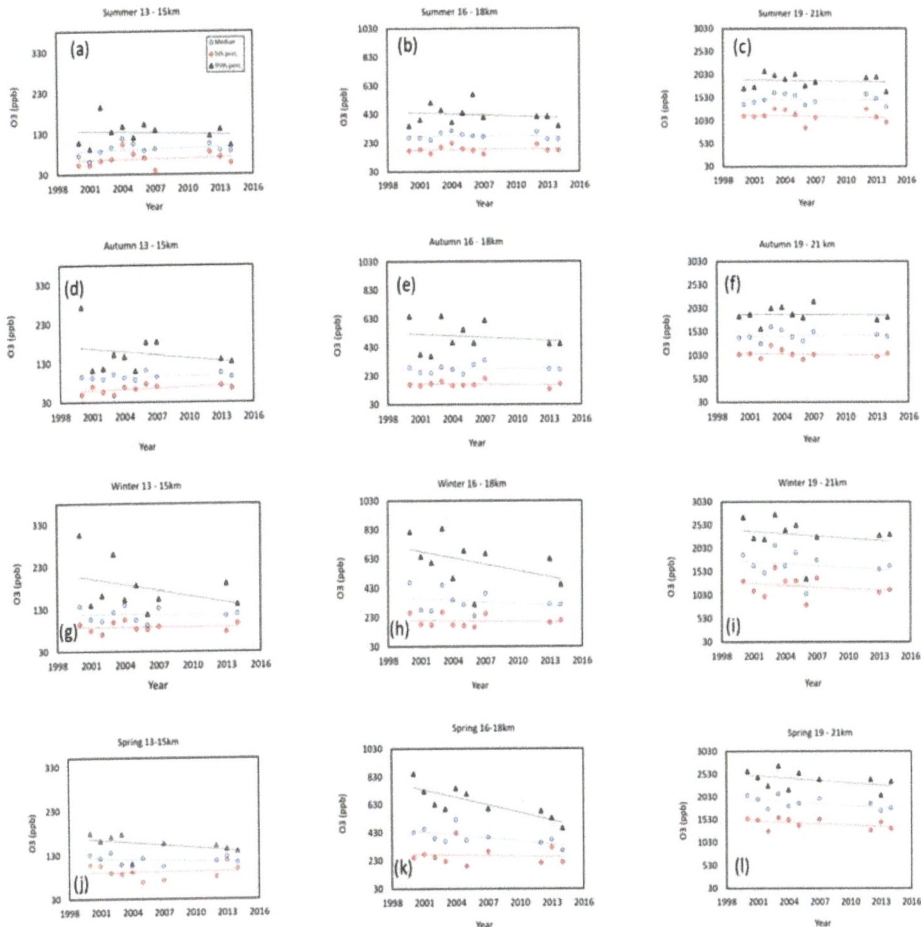

Figure 13. Ozone time series (2000–2015) of the 5th (red), median (blue) and 95th (black) percentile ozone mole fractions at different altitude, (**a**) summer 13–15 km, (**b**) summer 16–18 km and (**c**) summer 19–21 km, (**d**) autumn 13–15 km, (**e**) autumn 16–18 km, (**f**) autumn 19–21 km, (**g**) winter 13–15 km, (**h**) winter 16–18 km, (**i**) winter 19–21 km, (**j**) spring 13–15 km, (**k**) spring 16–18 km and (**l**) spring 19–21 km. Dashed lines represent the linear fit for each time series.

10–12 km layer: there was a negative change observed in autumn for the medians (−0.03 ± 0.81 ppb/year) and for the 5th percentiles in summer (−0.07 ± 0.66 ppb/year), autumn (−0.13 ± 0.61ppb/year) and winter (−1.00 ± 0.85 ppb/year). Similarly, there was a negative change observed in winter (−0.96 ± 2.03 ppb/year) and spring (−0.36 ± 1.19 ppb/year) for the 95th percentiles. Whilst on the other hand, a positive change was observed in summer (1.09 ± 0.72 ppb/year) and autumn (1.17 ± 1.04 ppb/year) for the 95th percentiles.

13–15 km layer: with the exception of winter (−0.18 ± 1.34 ppb/year) and spring (−0.50 ± 0.88 ppb/year), a positive change was observed in summer (0.91 ± 1.26ppb/year) and autumn (0.62 ± 0.67 ppb/year) for the medians. Similarly, a positive change was also observed in summer (0.59 ± 1.47 ppb/year), autumn (1.02 ± 0.81 ppb/year), winter (0.03 ± 0.93 ppb/year) and spring (0.24 ± 1.31 ppb/year) for the 5th percentiles. Contrasting with the 10–12 km layer, the 95th percentiles yielded negative change in all of the seasons.

Table 4. Ozone statistical summary at different altitudes and seasons.

Altitude	Median [ppb/year]	5th Percentile [ppb/year]	95th Percentile [ppb/year]
(a). Summer (±indicates the standard deviation)			
7–9 km	0.53 ± 0.40	0.60 ± 0.58	0.23 ± 0.78
10–12 km	0.48 ± 0.49	−0.07 ± 0.66	1.09 ± 0.72
13–15 km	0.91 ± 1.26	0.59 ± 1.47	−0.59 ± 2.36
16–18 km	−0.28 ± 1.84	0.56 ± 1.96	−2.08 ± 5.58
19–21 km	−1.28 ± 8.96	−4.69 ± 10.01	−3.39 ± 12.09
22–24 km	−21.78 ± 17.53	−25.66 ± 29.31	−18.00 ± 14.98
(b). Autumn (±indicates the standard deviation)			
7–9 km	−0.11 ± 0.53	0.40 ± 0.34	0.68 ± 0.79
10–12 km	−0.03 ± 0.81	−0.13 ± 0.61	1.17 ± 1.04
13–15 km	0.62 ± 0.67	1.02 ± 0.81	−2.50 ± 4.07
16–18 km	0.93 ± 2.53	−0.87 ± 1.69	−3.77 ± 8.74
19–21 km	1.40 ± 8.90	−2.94 ± 1.06	−0.99 ± 13.61
22–24 km	−12.54 ± 19.67	−13.77 ± 20.90	−13.59 ± 19.54
(c). Winter (±indicates the standard deviation)			
7–9 km	0.85 ± 0.72	0.06 ± 0.56	1.14 ± 0.94
10–12 km	0.14 ± 0.61	−1.00 ± 0.85	−0.96 ± 2.03
13–15 km	−0.18 ± 1.34	0.03 ± 0.93	−4.57 ± 4.89
16–18 km	−3.85 ± 6.26	−0.88 ± 3.31	−14.46 ± 13.17
19–21 km	−13.30 ± 23.48	−10.60 ± 18.75	−16.47 ± 31.56
22–24 km	−10.14±31.64	−25.50±34.14	−8.47 ± 33.79
(d). Spring (±indicates the standard deviation)			
7–9 km	0.04 ± 0.57	−0.30 ± 0.76	−0.54 ± 1.00
10–12 km	0.57 ± 0.42	0.90 ± 1.00	−0.36 ± 1.19
13–15 km	−0.50 ± 0.88	0.24 ± 1.31	−1.85 ± 1.80
16–18 km	−7.11 ± 4.99	−1.15 ± 5.73	−18.19 ± 9.57
19–21 km	−14.61 ± 10.94	−9.93 ± 9.50	−17.00 ± 15.99
22–24 km	−20.18 ± 18.49	−19.83 ± 25.19	−19.19 ± 18.97

16–18 km layer: with the exception of autumn (0.93 ± 2.53 ppb/year), negative changes were observed in summer (−0.28 ± 1.84 ppb/year), winter (−3.85 ± 6.26 ppb/year) and spring (−7.11 ± 4.99 ppb/year) for the medians. Similarly, there was a negative change observed in autumn (−0.87 ± 1.69 ppb/year), winter (−0.88 ± 3.31 ppb/year) and spring (−1.15 ± 5.73 ppb/year) for the 5th percentiles except in summer (0.56 ± 1.96 ppb/year). The 95th percentiles again showed negative changes in all of the seasons. Standard deviations of more than 1.5 ppb were observed in all seasons.

19–21 km layer: with the exception of autumn (1.40 ± 8.90 ppb/year), negative changes were observed for the medians in summer (−1.28±8.96 ppb/year), winter (−13.30 ± 23.48 ppb/year) and spring (−14.61 ± 10.94ppb/year). In addition to this, negative changes were also observed in all seasons for the 5th and 95th percentiles. With the exception of 5th percentiles in autumn, standard deviations in excess of 8.0 ppb were calculated in this layer with more variation observed during winter and spring.

22–24 km layer: There was a negative change observed in all seasons for the medians, 5th and 95th percentiles. A standard deviation of more 14.0 ppb was calculated in this layer with more variation in winter and spring. The observed high standard deviations suggest greater significance of changes within this layer.

4. Summary and Conclusions

This study examined Irene O_3 profile data from 2000–2015 in order to identify high O_3 events and to study O_3 decline at different altitudes of the stratosphere. Monthly 90th percentile composites were used as a threshold to identify high O_3 events. Furthermore, PV charts at isobaric level were used to identify high PV air mass of stratospheric origin (more than 2 PVU). Based on the observations, high O_3 events were found to occur in all seasons. However, they were most prevalent in winter and spring. The results showed that high O_3 of stratospheric origin can propagate down to 7 km over Irene.

However, very few events were found to reach this altitude. The majority of events occurred between 9 km and 10 km from the earth surface. Based on the results obtained from the PV charts, high PV values of approximately 3 PVU were observed over Irene.

Furthermore, O_3 data was grouped into three categories: 2000–2003, 2004–20007 and 2012–2015 to investigate possible long-term changes using monthly 5th percentile composites, monthly 95th percentile composites and monthly median composites. Troposphere and stratosphere O_3 vertical profiles were generated for the three datasets (2000–2003, 2004–20007 and 2012–2015). Based on the vertical profile graphs, it was noted that the maximum standard deviation occurred in altitudes closer to the tropopause (approximately 17 km). This could be related to STE and other dynamic changes occurring in the tropopause region.

The annual changes presented in Table 3 show an O_3 decline at 13–15 km for the 95th percentiles (−2.38 ± 3.28 ppb/year) while median and 5th percentile O_3 decline started at the 16–18 km layer. A maximum decline was observed at 22–24 km for the medians (−16.16 ± 21.83 ppb/year), 5th (−21.19 ± 27.39 ppb/year) and 95th percentile (−14.81 ± 21.82 ppb/year). High O_3 decline was observed at 19–21 km and 22–24 km in all seasons. The 95th percentiles show O_3 decline at 13–15 km. While O_3 decline was observed in winter and spring at 10–12 km layer for medians and 95th percentiles. In conclusion, high O_3 of stratospheric origin can occasionally reach down as low as 7 km above Irene. However, 68.8% of these events were observed within the 9 km to 10 km region.

PV charts proved a very useful tool and showed the propagation of stratospheric air masses to the troposphere as further evidence of stratosphere O_3 intrusion for the selected high O_3 episodes in this study. These observations seem to indicate that STE events which are observed in over Irene are strongly driven by the dynamics of the Southern Hemisphere polar vortex.

O_3 decline was observed mainly in the lower stratosphere (16–28 km). However, it was more dominant in winter and spring, while few events were observed in summer and autumn. Contrary to this, O_3 increase was observed in the lower troposphere. These observations of O_3 increase in the lower troposphere are in line with literature reports and were associated with an increase pollution in the lower troposphere.

Author Contributions: Conceptualization, T.M. and N.M.; methodology, T.M.; validation, T.M.; formal analysis, T.M.; N.M. and N.B.; investigation, T.M. and N.M.; resources, T.M.; writing—original draft preparation, T.M.; writing—review and editing, T.M.; N.M.; V.S.; G.C.; C.L.; visualization, T.M.; N.M.; and V.S. All authors have read and agreed to the published version of the manuscript.

Funding: This research was funded jointly by the CNRS (Centre National de la Recherche Scientifique) and the NRF (National Research Foundation) in the framework of the LIA ARSAIO and by the South Africa/France PROTEA Program (project No 42470VA).

Acknowledgments: We thank South African Weather Service and ECMWF for access to the dataset. Our extended gratitude goes to Hassan Bencherif for his meaningful support. The ozonesonde data used here was obtained from the SHADOZ website https://tropo.gsfc.nasa.gov/shadoz/.

Conflicts of Interest: The authors declare no conflict of interest.

References

1. Bekki, S.; Lefevre, F. Stratospheric ozone: History and concepts and interactions with climate. *Eur. Phys. J. Conf.* **2009**, *1*, 113–136. [CrossRef]
2. Farman, J.C.; Gardiner, B.G.; Shanklin, J.D. Large losses of total ozone in Antarctica reveal seasonal CLOX/NOX interaction. *Nature* **1985**, *315*, 207–210. [CrossRef]
3. Morrisette, P.M. The evolution of policy responses to tratospheric ozone depletion. *Nat. Res. J.* **1989**, *29*, 793–820.
4. Rowland, F.S.; Molina, M.J. Chlorofluoromethanes in environment. *Rev. Geophys.* **1975**, *13*, 1–35. [CrossRef]
5. Mäder, J.A.; Staehelin, J.; Peter, T.; Brunner, D.; Rieder, H.E.; Stahel, W.A. Evidence for the effectiveness of the Montreal Protocol to protect the ozone layer. *Atmos. Chem. Phys.* **2010**, *10*, 12161–12171. [CrossRef]
6. Scientific assessment of ozone depletion: 2010. In *Global Ozone Research and Monitoring Project-Report No. 52*; World Meteorological Organization (WMO): Geneva, Switzerland, 2011; p. 516.

7. Bodeker, G.E.; Scott, J.C.; Kresher, K.; McKenzie, R.L. Global Ozone Trends in Potential Vorticity Coordinates Using TOMS and GOMEE Inter-Compared Against the Dobson network. *J. Geophys. Res.* **2001**, *106*, 23029–23042. [CrossRef]
8. Scientific assessment of ozone depletion (2014). In *Global Ozone Research and Monitoring Project Report*; World Meteorological Organization (WMO): Geneva, Switzerland, 2014; p. 416.
9. Ball, W.T.; Alsing, J.; Mortlock, D.J.; Rozanov, E.V.; Tummon, F.; Haigh, J.D. Reconciling differences in stratospheric ozone composites. *Atmos. Chem. Phys.* **2017**, *17*, 12269–12302. [CrossRef]
10. Ball, W.T.; Alsing, J.; Mortlock, D.J.; Staehelin, J.; Haigh, J.D.; Peter, T.; Tummon, F.; Stübi, R.; Stenke, A.; Anderson, J.; et al. Evidence for a continuous decline in lower stratospheric ozone offsetting ozone layer recovery. *Atmos. Chem. Phys.* **2018**, *18*, 1379–1394. [CrossRef]
11. Seinfeld, J.H.; Pandis, S.N. *From Air Pollution to Climate Change*; John Wiley and Sons: New York, NY, USA, 1998; p. 1326.
12. Haagen-Smit, A.J. Chemistry and physiology of Los Angeles smog. *Ind. Eng. Chem.* **1952**, *44*, 1342–1346. [CrossRef]
13. El Amraoui, L.; Atti´e, J.L.; Semane, N.; Claeyman, M.; Peuch, V.-H.; Warner, J.; Ricaud, P.; Cammas, J.-P.; Piacentini, A.; Josse, B.; et al. Midlatitude stratosphere—Troposphere exchange as diagnosed by MLS O3 and MOPITT CO assimilated fields. *Atmos. Chem. Phys.* **2010**, *10*, 2175–2194. [CrossRef]
14. Diab, R.D.; Thompson, A.M.; Mari, K.; Ramsay, L.; Coetzee, G.J.R. Tropospheric ozone climatology over Irene, South Africa from 1990 to 1994 and 1998 to 2000. *J. Geophys. Res.* **2004**, *109*, JD00479. [CrossRef]
15. Ziemke, J.R.; Chandra, S.; Duncan, B.N.; Froidevaux, L.; Bhartia, P.K.; Levelt, P.F.; Waters, J.W. Tropospheric ozone determined from Aura OMI and MLS: Evaluation of measurements and comparison with the Global Modeling Initiative's Chemical Transport Model. *J. Geophys. Res.-Atmos.* **2006**, *111*, D19303. [CrossRef]
16. Ziemke, J.R.; Chandra, S.; Labow, G.J.; Bhartia, P.K.; Fridevaux, L.; Witte, J.C. A global Climatology of Tropospheric and Stratospheric Ozone Derived from AURA OMI and MLS Measurements. *Atmos. Chem. Phys.* **2011**, *11*, 9237–9251. [CrossRef]
17. Sivakumar, V.; Bencherif, H.; Begue, N.; Thompson, A.M. Tropopause characteristics and variability from 11 years of SHADOZ Observations in the Southern Tropics and Subtropics. *J. Appl. Meteorol. Clim.* **2011**, *50*, 1403–1416. [CrossRef]
18. Thompson, A.M.; Miller, S.K.; Tilmes, S.; Kollonige, D.W.; Witte, J.C.; Oltmans, S.J.; Johnson, B.J.; Fujiwara, M.; Schmidlin, F.J.; Coetzee, G.J.R.; et al. Southern Hemisphere Additional Ozonesondes (SHADOZ) ozone climatology (2005–2009): Tropospheric and tropical tropopause layer (TTL) profiles with comparisons to OMI–based ozone products. *J. Geophys. Res.-Atmos.* **2012**, *17*, D23301. [CrossRef]
19. Poulida, O.; Dickerson, R.R.; Heymsfield, A. Stratosphere—Troposphere exchange in a midlatitude mesoscale con- vective complex. 1. Observations. *J. Geophys. Res.* **1996**, *101*, 6823–6836. [CrossRef]
20. Mulumba, J.-P.; Sivakumar, V.; Afullo, T.J.O. Modeling Tropospheric ozone climatology over Irene (South Africa) using retrieved remote sensing and ground-based measured data. *J. Geosci. Remote Sens.* **2015**. [CrossRef]
21. Ndarana, T.; Waugh, D.W. The link between cut-off lows and Rossby wave breaking in the Southern Hemisphere. *Q. J. R. Meteorol. Soc.* **2010**, *136*, 869–885. [CrossRef]
22. Sun, L.; Chen, G.; Robinson, W.A. The Role of Stratospheric Polar Vortex Breakdown in Southern Hemisphere Climate Trends. *BAMS Meteol. Soc.* **2014**. [CrossRef]
23. Garfinkel, I.; Hartmann, D.L. The influence of the quasi-biennial oscillation on the troposphere in winter in a Hierachy of Models. Part 1: Simplified dry GCMs. *AMS Meteol. Soc.* **2011**. [CrossRef]
24. Holton, J.R.; Tan, H.C. The influence of the equatorial quasi–biennial oscillation on the global circulation at 50 mb. *J. Atmos. Sci.* **1980**, *37*, 2200–2208. [CrossRef]
25. Hamilton, K. Effects of an imposed quasi-biennial oscillation in a comprehensive troposphere—Stratosphere—Mesosphere general circulation model. *J. Atmos. Sci.* **1998**, *55*, 2393–2418. [CrossRef]
26. Baldwin, M.P.; Dunkerton, T.J. Stratospheric harbingers of anomalous weather regimes. *Science* **2001**, *294*, 581–584. [CrossRef] [PubMed]
27. Waugh, D.W.; Polvani, L.M. Climatology of intrusions into the tropical upper troposphere. *Geophys. Res. Lett.* **2000**, *27*, 3857–3860. [CrossRef]

28. Semane, N.; Bencherif, H.; Morel, B.; Hauchecorne, A.; Diab, R.D. An unusual stratospheric ozone decrease in the Southern Hemisphere subtropics linked to isentropic air-mass transport as observed over Irene (25.5 S, 28.1 E) in mid-May 2002. *Atmos. Chem. Phys.* **2006**, *6*, 1927–1936. [CrossRef]
29. Bencherif, H.; Amraoui, L.E.; Kirgis, G.; Leclair De Bellevue, J.; Hauchecorne, A.; Mzé, N.; Portafaix, T.; Pazmino, A.; Goutail, F. Analysis of a rapid increase of stratospheric ozone during late austral summer 2008 over Kerguelen (49.4 S, 70.3 E). *Atmos. Chem. Phys.* **2011**, *11*, 363–373. [CrossRef]
30. Orte, P.F.; Wolfram, E.; Salvador, J.; Mizuno, A.; Bègue, N.; Bencherif, H.; Bali, J.L.; d'Elia, R.; Pazmino, A.; Godin-Beekmann, S.; et al. Analysis of a southern sub-polar short-term ozone variation event using a millimetre-wave radiometer. *Ann. Geophys.* **2019**, *37*, 613–629. [CrossRef]
31. Greenslade, J.W.; Alexander, S.P.; Schofield, R.; Fisher, J.A.; Klekociuk, A.K. Stratospheric ozone intrusion and their impacts on tropospheric ozone. *Atmos. Chem. Phys.* **2017**. [CrossRef]
32. Helmig, D.; Oltmans, S.J.; Carlson, D.; Lamarque, J.-F.; Jones, A.; Labuschagne, C.; Anlauf, K.; Hayden, K. A review of surface ozone in the polar regions. *Atmos. Environ.* **2007**, *41*, 5138–5161. [CrossRef]
33. Cooper, O.R.; Parrish, D.D.; Ziemke, J.; Balashov, N.V.M.; Cupeiro, M.; Galbally, I.E.; Gilge, S.; Horowitz, L.; Jensen, N.R.; Lamarque, J.-F.; et al. Global distribution and trends of tropospheric ozone: An observation-based review. *Elem.: Sci. Anthr.* **2014**, 2. [CrossRef]
34. Oltmans, S.J.; Lefohn, A.S.; Shadwick, D.; Harris, J.M.; Scheel, H.E.; Galbally, I.; Tarasick, D.W.; Johnson, B.J.; Brunke, E.-G.; Claude, H.; et al. Recent tropospheric ozone changes—A pattern dominated by slow or no growth. *Atmos. Environ.* **2013**, *67*, 331–351. [CrossRef]
35. Lin, M.; Horowitz, L.W.; Cooper, O.R.; Tarasick, D.; Conley, S.; Iraci, L.T.; Johnson, B.; Leblanc, T.; Petropavlovskikh, I.; Yates, E.L. Revisiting the evidence of increasing springtime ozone mixing ratios in the free troposphere over western North America. *Geophys. Res. Lett.* **2015**, 8719–8728. [CrossRef]
36. Xu, W.; Lin, W.; Xu, X.; Tang, J.; Huang, J.; Wu, H.; Zhang, X. Long-term trends of surface ozone and its influencing factors at the Mt Waliguan GAW station, China—Part 1: Overall trends and characteristics. *Atmos. Chem. Phys.* **2016**, *16*, 6191–6205. [CrossRef]
37. Granados–Muñoz, M.J.; Leblanc, T. Tropospheric ozone seasonal and long–term variability as seen by LIDAR and surface measurements at the JPL–Table Mountain Facility, California. *Atmos. Chem. Phys.* **2016**, *16*, 9299–9319. [CrossRef]
38. Sivakumar, V.; Ogunniyi, J. Ozone climatology and variability over Irene, South Africa determined by ground based and satellite observations. Part 1: Vertical variations in the troposphere and stratosphere. *Atmosfera* **2017**, *30*, 337–353. [CrossRef]
39. Holton, J.R.; Hayenes, P.H.; McInyre, M.E.; Douglass, R.A.; Roodand, R.B.; Pfister, L. Stratosphere—Troposphere exchange. *Rev. Geophys.* **1995**, *33*, 403–439. [CrossRef]
40. National Aeronautics and Space Administration Goddard Space Flight Center. Available online: https://tropo.gsfc.nasa.gov/shadoz/ (accessed on 20 April 2016).
41. Newell, R.E.; Browell, E.V.; Davis, D.D.; Liu, S.C. Western Pacific tropospheric ozone and potential vorticity: Implications for Asian pollution. *Geophys. Res. Lett.* **1997**, *24*, 2733–2736. [CrossRef]
42. World Meteorological Organization (WMO). *Atmospheric Ozone 1985, Vol. I.*; World Meteorological Organization: Geneva, Switzerland, 1986; p. 478.
43. Shapiro, M.A. Turbulent mixing within tropopause folds as a mechanism for the exchange of chemical constituents between the stratosphere and the troposphere. *J. Atmos. Sci.* **1980**, *37*, 994–1004. [CrossRef]
44. National Aeronautics and Space Administration Goddard Institute for Space Studies Website. Available online: https://www.giss.nasa.gov/tools/panoply (accessed on 22 November 2018).
45. Tang, Q.; Prather, M.J. Correlating tropospheric column ozone with tropopause folds: The Aura-OMI satellite data. *Atmos. Chem. Phys.* **2010**, *10*, 9681–9688. [CrossRef]
46. Thompson, A.M.; Balashov, J.C.; Coetzee, J.R.C.; Thouret, V.; Posny, F. Tropospheric ozone increases over the southern Africa region: Bellwether for rapid growth in Southern Hemisphere pollution? *Atmos. Chem. Phys.* **2014**, *14*, 9855–9869. [CrossRef]
47. Clain, G.; Baray, J.L.; Delmas, R.; Keckhut, P.; Cammas, J.P. A lagrangian approach to analyse the tropospheric ozone climatology in the tropics: Climatology of stratosphere-troposphere exchange at Reunion Island. *Atmos. Environ.* **2010**, *44*, 968–975. [CrossRef]
48. Stan, C.; Randall, D.A. Potential vorticity as Meridonal coordinate. *BAM Meteol. Soc.* **2007**, *64*, 23029–23042.

49. Hoang, L.P.; Reeder, M.J.; Berry, G.J.; Schwendike, J. Coherent Potential Vorticity Maxima and Their Relationship to Extreme Summer Rainfall in the Australian and North African Tropics. *J. South. Hemisph. Earth* **2016**, *66*, 424–441. [CrossRef]
50. Waugh, D.W.; Polvani, L.M. *Stratospheric Polar Vortices, in the Stratosphere: Dynamics, Transport, and Chemistry*; Polvani, L.M., Sobel, A.H., Waugh, D.W., Eds.; AGU: Washington, DC, USA, 2013; Volume 190, pp. 43–57.
51. Tyson, P.D.; Preston-Whyte, R.A. *The Weather and Climate of Southern Africa*; Oxford Univ. Press: New York, NY, USA, 2000.
52. Hauchecorne, A.; Godin, S.; Marchand, M.; Heese, B.; Souprayen, C. Quantification of the transport of chemical constituents from the polar vortex to midlatitudes in the lower stratosphere using the high-resolution advection model MIMOSA and effective diffusivity. *J. Geophys. Res.: Atmos.* **2002**, *107*, SOL-32. [CrossRef]
53. Dowdy, A.J.; Vincent, R.A.; Murphy, D.J.; Tsutsumi, M.; Riggin, D.M.; Jarvis, M.J. The large-scale dynamics of the mesosphere—lower thermosphere during the Southern Hemisphere stratospheric warming of 2002. *Geophys. Res. Lett.* **2004**, *31*. [CrossRef]
54. Mbatha, N.; Sivakumar, V.; Malinga, S.B.; Bencherif, H.; Pillay, S.R. Study on the impact of sudden stratosphere warming in the upper mesosphere-lower thermosphere regions using satellite and HF radar measurements. *Atmos. Chem. Phys.* **2010**, *10*, 3397–3404. [CrossRef]
55. Baray, J.L.; Ancellet, G.; Taupin, F.G.; Bessafi, M.; Baldy, S.; Keckhut, P. Subtropical tropopause break as a possible stratospheric source of ozone in the tropical troposphere. *J. Atmos. Sol. Terr. Phys.* **1998**, *60*, 27–36. [CrossRef]
56. Thompson, A.M.; Witte, J.C.; Freiman, M.T.; Phahlane, N.A.; Coetzee, G.J.R. Lusaka, Zambia, during SAFARI–2000: Convergence of local and imported ozone pollution. *Geophys. Res. Lett.* **2002**, *29*. [CrossRef]
57. Bourassa, A.E.; Roth, C.Z.; Zawada, D.J.; Rieger, L.A.; McLinden, C.A.; Degenstein, D.A. Drift corrected Odin-OSIRIS ozone product: Algorithm and updated stratospheric ozone trends. *Atmos. Meas. Tech. Discuss.* **2017**. [CrossRef]
58. Sofieva, V.; Kyrölä, E.; Laine, M.; Tamminen, J.; Degenstein, D.; Bourassa, A.; Roth, C.; Zawada, D.; Weber, M.; Rozanov, A.; et al. Merged SAGE II, Ozone_cci and OMPS ozone profiles dataset and evaluation of ozone trends in the stratosphere. *Atmos. Chem. Phys. Discuss.* **2017**. [CrossRef]
59. Steinbrecht, W.; Froidevaux, L.; Fuller, R.; Wang, R.; Anderson, J.; Roth, C.; Bourassa, A.E.; Degenstein, D.A.; Damadeo, R.; Zawodny, J.M.; et al. An update on ozone profile trends for the period 2000 to 2016. *Atmos. Chem. Phys. Discuss.* **2017**, 1–24. [CrossRef]
60. WMO: Scientific assessment of ozone depletion: 2018. In *Global Ozone Research and Monitoring Project-Report*; World Meteorological Organization: Geneva, Switzerland, 2018; p. 588.
61. Petropavlovskikh, I.; Godin-Beekmann, S.; Hubert, D.; Damadeo, R.; Hassler, B.; Sofieva, V. SPARC/IO3C/GAW, 2019: SPARC/IO3C/GAW Report on Long-Term Ozone Trends and Uncertainties in the Stratosphere. Available online: https://www.sparc-climate.org/publications/sparc-reports/sparc-report-no-9/ (accessed on 16 May 2019).
62. Wargan, K.; Orbe, C.; Pawson, S.; Ziemke, J.R.; Oman, L.D.; Olsen, M.A.; Coy, L.; Knowland, E.K. Recent decline in extratropical lower stratospheric ozone attributed to circulation changes. *Geophys. Res. Lett.* **2018**, *45*. [CrossRef]

Article

Ozone Trends in the United Kingdom over the Last 30 Years

Florencia M. R. Diaz [1,2], M. Anwar H. Khan [1], Beth M. A. Shallcross [3], Esther D. G. Shallcross [3], Ulrich Vogt [2] and Dudley E. Shallcross [1,*]

1 School of Chemistry, University of Bristol, Bristol BS8 1TS, UK; flor.ramdi@gmail.com (F.M.R.D.); anwar.khan@bristol.ac.uk (M.A.H.K.)
2 Institute of Combustion and Power Plant Technology, University of Stuttgart, D-70569 Stuttgart, Germany; ulrich.vogt@ifk.uni-stuttgart.de
3 School of Pharmacy, The University of Manchester, Oxford Road, Manchester M13 9PL, UK; bma.shallcross@gmail.com (B.M.A.S.); esther.shallcross@gmail.com (E.D.G.S.)
* Correspondence: d.e.shallcross@bris.ac.uk; Tel.: +44-117-9287796

Received: 22 April 2020; Accepted: 19 May 2020; Published: 21 May 2020

Abstract: Previous work regarding the behaviour of ozone surface concentrations over many years in the United Kingdom had predicted that the frequency and severity of ozone episodes would become less marked in the future as a response to environmental regulations. The aim of this study is to extend these studies and compare the results with their predictions. The ozone data of 13 rural and six urban sites in the UK collected from the Department for Environment, Food and Rural Affairs over a period from 1992 to mid-2019 were used to investigate this behaviour. The yearly ozone exceedances (the number of hours that the ozone concentration exceeded the 50 ppbv limit) in the United Kingdom were found to have decreased over the last 30 years regardless of the type of site (rural or urban), showing that the adopted emission controls have so far been successful in the abatement of pollutant emissions. In the past three decades, the highest numbers of exceedances were reached in May regardless of the type of site. Furthermore, these episodes have become less frequent and less severe in recent years. In fact, the number of hours of exceedance is lower than that in previous decades, and it is almost constant throughout the week.

Keywords: ozone exceedance; urban site; rural site; human health

1. Introduction

There is an increasing interest in the study of the behaviour of tropospheric ozone concentrations over the past 30 years due to its role as a greenhouse gas with an estimated globally averaged radiative forcing of 0.4 ± 0.2 Wm^{-2}, as a component of smog and as a primary tropospheric source of the hydroxyl radical (OH), which is a dominant tropospheric oxidant determining the lifetime of trace gases [1,2]. As a pollutant, ozone is corrosive and severely damaging to plants, trees and even buildings [3–5]. It has also been shown to have serious health effects on humans, particularly affecting the respiratory, cardiovascular and central nervous systems [6,7]. For example, it can cause irritation; it reduces the function of the lungs and promotes susceptibility to respiratory infections [8].

Tropospheric ozone is not directly emitted; it is formed as a secondary pollutant in the boundary layer by the reaction of primary pollutants (e.g., nitrogen dioxide and hydrocarbons) in the presence of sunlight. Because of this, its abatement depends on various parameters related to the emissions of these pollutants. Furthermore, ozone is transboundary [9], i.e., emissions from distant locations can contribute to its formation at a specific site; regulating the concentration of this pollutant requires international efforts. Ozone formation depends on the VOC–NO_X ratio [10]. In urban areas, ozone concentrations are expected to rise due to its formation through NO_X photochemistry, which originates from road traffic exhaust, particularly primary diesel-fuelled vehicles at low speeds [11]. Ozone extremes have been found to have decreased during the last decade due to the substantial reduction of NO_X emissions in response to protocols [12–14]. Rural background ozone concentrations have been found to be higher than those in urban backgrounds [15] due to two well established reasons, one being the (northern hemispheric) ozone baseline and the second being the presence of more NO_X in urban sites actively scavenging ozone. A decrease in these pollutants would then lead urban sites to start behaving like their rural counterparts, and ozone concentrations would increase over the years as was found in [16].

Given that the formation of ozone depends on the concentration of the primary pollutants, a decrease in the concentration of pollutants like NO_X causes ozone to decrease under the NO_X-limited regime, while in cities (under the VOC-limited regime), a decrease in NO_X leads to an increase in ozone levels. The Tropospheric Ozone Assessment Report (TOAR) showed that there is no clear global pattern for surface ozone changes since 2000, with increasing and decreasing trends in both polluted and remote sites [17]. Environmental policies have already been established to decrease the emissions of these pollutants. In Europe, for example, the Directive on Ambient Air Quality and Cleaner Air (2008/50/EC), adopted in 2008, aims to assess and manage the concentration thresholds of pollutants by setting limits [18]. If these limits are exceeded then authorities are required to implement plans to decrease these concentrations, as well as establish sanctions. These limits have also been observed by individual countries, and local objectives have been established. In the case of ozone, the objective limit in Europe is an 8 h mean of 60 ppbv (not to be exceeded more than 25 times a year, averaged over three years) [19], whereas in the United Kingdom, this limit is 50 ppbv (not to be exceeded more than 10 times a year) [20].

In order to understand what these policies have achieved in the reduction of the levels of the primary pollutants over the years, studies on ozone concentration trends in the United Kingdom have already been made [16,21–24]. This study is an update to these previous studies, addressing a more recent period and focusing on specific measuring sites throughout the UK. This study uses the freely available air quality data from the Department for Environment, Food and Rural Affairs (Defra) on selected rural and urban sites to address ozone exposure and understand its impact on human health. The aim of this study is to compare the results obtained in this work with the predictions made by the previous studies and determine if the implemented policies have succeeded or need to be amended. This study is thus a benchmark to indicate whether these policies have accomplished their objectives to date, or if pollution in the UK has worsened over the past three decades. Therefore, we analysed the ozone exceedances over the time period of 1992 to 2019 at 13 rural and six urban sites in the United Kingdom using the Defra ozone archive data. The decadal, yearly, monthly, daily and hourly variations of ozone exceedances are discussed in the study. We also analysed the magnitude and frequency of the most severe ozone episodes in rural and urban sites over the last 30 years.

2. Methodology

The source of the data for several measurement sites distributed throughout the UK was the Department for Environment, Food and Rural Affairs' (Defra's) Automatic Urban and Rural Network (AURN) data archive [25]. AURN is the UK's largest monitoring network comprising 150 sites and reporting hourly measurements of NO_X, SO_2, ozone, CO and particulate matter. We selected 13 rural and six urban sites in this study, which was based on the availability of the ozone data covering the last three decades (Figure 1). All these selected sites have an annual coverage of > 80% validated hourly data for ozone over the period from January 1992 to June 2019.

Ozone and NO_X measurements were carried out by UV photometry and chemiluminescence, respectively, following the guidelines of the European Committee for Standardisation [26]. Data were validated on an ongoing basis by manual review to exclude any errors due to instrument malfunctions or faulty calibrations [27]. The ozone and NO_X data (in $\mu g/m^3$ at 20 °C and 101.3 kPa) for each site were archived at the National Air Quality Information Archive [25]. The uncertainty (expressed as a

95% confidence level) of the measurement datasets and sites was around 15% [28]. The analysis was centred around finding the number of hours the ozone concentration exceeded the 50 ppbv limit (also known as an exceedance) in every selected site for a period from January 1992 to June 2019. These dates were chosen based on the date when urban sites began reporting ozone concentrations (rural reports go back as far as 1986) and the date when this study was first started. A linear regression method was used to estimate the trends in magnitude. Statistical significance is based on a $p < 0.001$ and the trends are reported with 95% confidence internals. Data are then analysed by type of site in order to find the frequency and magnitude of ozone episodes, as well as how trends have changed over the past three decades. Furthermore, in order to establish the causes for these pollution episodes, a meteorological back-trajectory model derived from the NOAA on-line trajectory service was used [29].

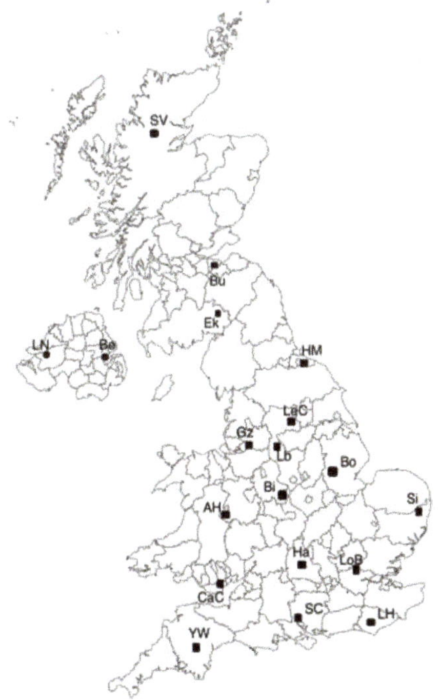

Rural	Urban
Aston Hill (AH)	Belfast Centre (Be)
Bottesford (Bo)	Birmingham Centre (Bi)
Bush Estate (Bu)	Cardiff Centre (CaC)
Eskdatemuir (Ek)	Leeds Centre (LeC)
Glazebury (Gz)	London Bloomsbury (LoB)
Harwell (Ha)	Southampton Centre (SC)
High Muffles (HM)	
Ladybower (Lb)	
Lough Navar (LN)	
Lullington Heath (LH)	
Sibton (Si)	
Strathvaich (SV)	
Yaner Wood (YW)	

Figure 1. Distribution of the measurement sites (both rural and urban) in the UK.

3. Results and Discussion

3.1. Three-Decadal Trend of Ozone Mixing Ratios

The yearly averaged maximum ozone mixing ratios for 13 rural sites and six urban sites were found to have decreased gradually over the last 30 years (Figure 2) at rates of 1.0 ppbv/y (1.2%/y, $p < 0.001$) and 0.68 ppbv/y (0.9%/y, $p = 0.002$), respectively (the individual sites' maximum ozone and average ozone trends can be found in Table S1). The year-to-year ozone exceedance variability is highly dependent on meteorology, which makes it hard to separate the trends caused by any other effects (e.g., reduced precursor emissions) [30,31]. However, an increasing trend in yearly average ozone was found at a rate of 0.13 ppb/y (0.5%/y, $p < 0.001$) for rural sites and 0.20 ppb/y (1.1%/y, $p < 0.001$) for urban sites (Figure 2). The increases in yearly average ozone can be explained by a decreasing trend in average NO_X mixing ratios at a rate of 0.22 ppb/y (3.7%/y, $p < 0.001$) for rural sites and 1.2 ppb/y (3.3%/y, $p < 0.001$) for urban sites, resulting in less NO_X scavenging [14,32,33], but the magnitude of the increasing trend at rural sites is smaller than those obtained at urban sites, due to the strong dependency of the concentrations of ozone on the northern hemispheric ozone baseline in rural areas [16]. Similar results for the decrease in maximum ozone concentration and increase in background ozone concentration are found in European sites for the period of 1995 to 2014 [34].

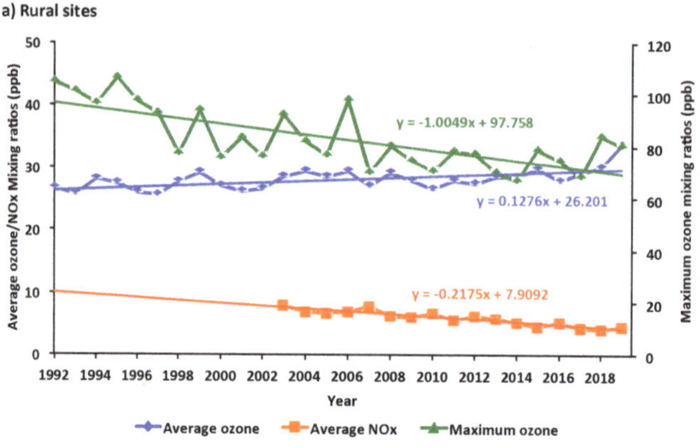

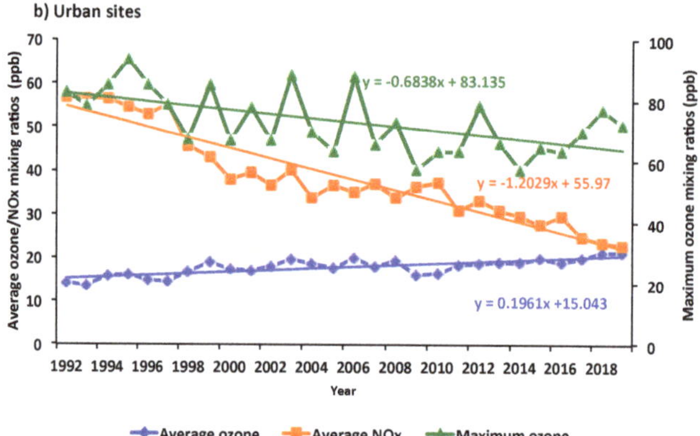

Figure 2. The mixing ratios of yearly average ozone, average NO_X and maximum ozone for (**a**) rural and (**b**) urban sites over the last 30 years. Note: The data points indicate the averages of all rural sites (**a**) and urban sites (**b**) of the yearly averaged data for the indicated time period. No available data for rural NO_X from 1992 to 2002. Trends are based on linear regression fitting with 95% confidence intervals and *p* values.

3.2. *Yearly Variation of Ozone Exceedances over the Three Decades*

The yearly total ozone exceedances for all rural sites show an overall decrease in the past three decades with the exceptions of the years 1992, 1995, 1999, 2003, 2006, 2008 and 2018 (Figure 3a). The weather conditions in the UK in the years 2003 [35,36] and 2006 [37] were particularly extreme, leading to a dramatic increase in summer pollution exceedances. The same applies for 1992 [38], 1995 [39] and 2018 [40], although the conditions were not as extreme in these years. This can be observed in Figure 3 where a high number of exceedances was reported for all these years. However, the high exceedances for 1999 and 2008 were not characterized by extreme weather conditions. The high exceedances in 1999 are due to a particularly long episode at the end of July, explained by the persistent anticyclonic conditions during this period preventing the bad air conditions from dissipating [23]. The high exceedances in 2008 were mainly a result of the influence of the site Strathvaich, which reached 1000 h of exceedance in 2008 (Figure 3a), a quarter of the overall hours of exceedance from every site for this year. Most of these exceedances were reported in the spring months (e.g., March, April and May) rather than summer months (e.g., June, July and August) when ozone episodes are expected to happen (see Figure S1b). There is a typical spring-time ozone maximum characteristic of the northern hemispheric baseline air [41]. This cycle, shown to achieve a maximum between March and April, is also found in Strathvaich, which is due to an unclassical behavior of the baseline cycle. This means that the high number of exceedances in 2008 is due to this baseline cycle rather than any influence from ozone precursor emissions, and thus, the maximum achieved in 2008 in rural sites is not due to any severe pollution episodes.

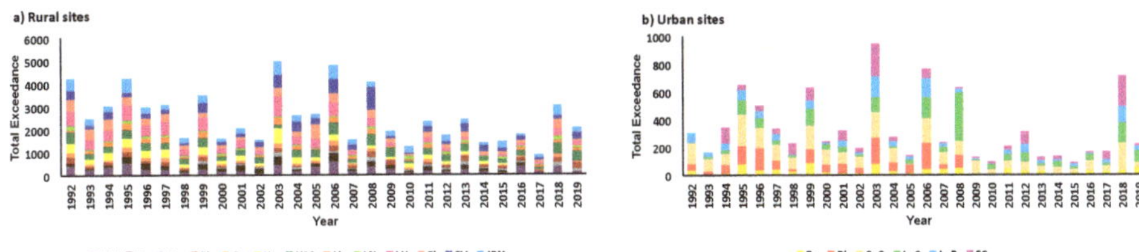

Figure 3. Yearly total exceedances of ozone in (**a**) rural and (**b**) urban sites over the last 30 years. Note that the total exceedance is the total number of hours the ozone concentration exceeded the 50 ppbv limit for (**a**) rural sites and (**b**) urban sites.

Similarly, for urban sites, the years with the highest exceedances were found to be 1995, 1999, 2003, 2006, 2008 and 2018 (see Figure 3b). In 1992, "peak years" were identified at rural sites, which had very few hours of exceedance at urban sites compared with the other years. The 2008 maximum in urban sites was also influenced by only one site, Leeds Centre, with over 350 h of exceedance (Figure 3b), more than half of the overall hours from each site for this particular year. In 2008, most of the total hours of exceedance in Leeds Centre were reported in May (see Figure S2b). During this month, two episodes took place, one lasting for 14 days, starting on the 5th and ending on the 18th, and another lasting for 8 days from the 20th to the 27th. The highest ozone mixing ratio was recorded on the 11th, reaching 86.6 ppbv. Both episodes were only seen at Cardiff Centre, but their durations were shorter, one lasting for 6 days and the other lasting for 3 days. Birmingham Centre also reported one of these episodes, one starting on the 5th and lasting for 6 days. The rest of the sites did not report either of the episodes; in fact, Belfast Centre, London Bloomsbury and Southampton Centre showed no consecutive exceedances in May. The high number of exceedances during this year is attributed to the episode that took place earlier in the month. A backward trajectory analysis was performed for the Birmingham Centre, Cardiff Centre and Leeds Centre sites in order to deduce the cause of this episode. The trajectories (see Figure S3), following a 96 h span on the day the highest mixing ratios were reached, pass through continental Europe prior to their arrival in the UK at 4 p.m. local time. Since the highest mixing ratio was recorded at Cardiff Centre, it can be assumed that this magnitude is due to the air parcel passing directly over London and transporting with it a high concentration of ozone precursors. It is likely that during this episode, there was an anticyclonic weather event and easterly flows, which transport pollution from continental Europe to the UK, favouring considerable ozone production [42].

The number of exceedances for the urban sites is much lower than that for the rural sites (Table 1) because in urban sites, NO_X emissions are more prevalent, which cause ozone scavenging, resulting in lower ozone mixing ratios. From the 1990s to the 2000s, the ozone exceedances increased. A decrease in NO_X from the 1990s to the 2000s (Figure 2) is likely to have led to less NO scavenging, and ozone is then supposed to have accumulated, increasing the number of hours of exceedance. However, from the 2000s to the 2010s, the ozone exceedances decreased, possibly because of the reduction in ozone precursor emissions (e.g., anthropogenic VOCs) caused by the 1999 Gothenburg protocol, which reduced VOC emissions in the UK by ~40% from the 2000s to the 2010s [43]. Overall, the exceedances for both urban and rural sites were decreased by 27% and 30%, respectively, from the 1990s to the 2010s.

Table 1. The average exceedances over the last three decades.

Type of Site	1990s	2000s	2010s	Overall
Rural	1941	2141	1412	5494
Urban	532	647	376	1554

Note: Average values accounting for the number of sites of each type.

3.3. Seasonal Trend Variation of Ozone Exceedances over the Three Decades

The rural and urban sites follow a seasonal exceedances trend, with a summer high and winter low throughout the three decades (Figure 4). The ozone episodes depend on the weather conditions and, above all else, sunlight, since its formation is led by photolysis. It is then logical to assume that in a period in which sunlight is constant and its duration is long, a larger amount of ozone would be formed and accumulated, increasing its concentration. Furthermore, ozone episodes are linked with summer anticyclonic weather, leading to warm and sunny conditions with low wind speeds, lowering the dissipation rate of pollutants. However, this trend is not the same for each decade. The seasonal exceedances have all tended to peak in May, but the way in which the trend emerges is different throughout the years. For example, in the 1990s and 2000s, the exceedances dropped gradually after reaching the May maximum, but in the 2010s, the exceedances dropped to a minimum in June, increased in July and dropped again in August (Figure 4). These discrepancies can be explained by the variable weather conditions from year to year. For example, the June minimum in the 2010s decade is solely explained by the different weather conditions; among the summer months, June is characterized by lower pressure and temperature, and higher rainfall in the UK [44]. From 2011 to 2016, the weather was consistent, with frequent rain and low temperatures [45–50]. However, in 2017, a hot spell was recorded in mid-June, with temperatures reaching up to 28 °C towards the south east of England [51]. In 2018, a particularly heavy rainfall event was recorded resulting from Storm Hector [52], and in 2019, rainfall, around 2.5 times heavier than average, was recorded towards the East Midlands [53]. It is well established that rainfall and elevated wind speeds have a cleaning effect on pollution emissions [54], and thus, heavy rainfall would result in fewer ozone exceedances.

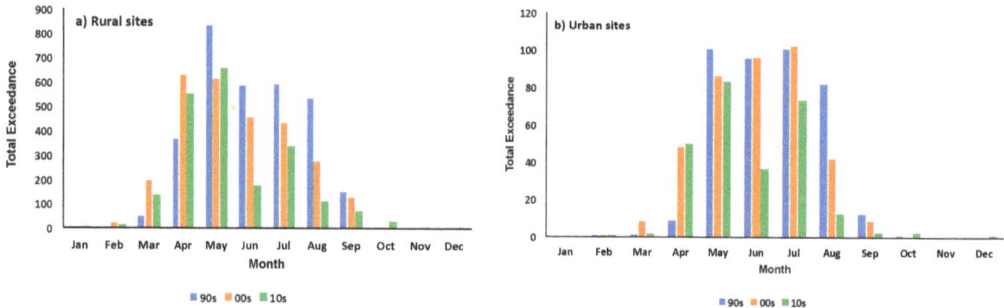

Figure 4. Seasonal total ozone exceedances in (**a**) rural and (**b**) urban sites averaged for the decades of the 1990s, 2000s and 2010s. Total exceedance is calculated as the total number of hours at an ozone concentration ≥ 50 ppbv for each month in a year, yearly averaged for each decade.

3.4. Daily Variations of Ozone Exceedances over the Three Decades

The diurnal plots for rural and urban sites (Figure 5) show that the highest ozone exceedances are found to have been reached at around 4 p.m. local time (mid-afternoon) throughout the three decades, and the lowest ozone exceedances were reported in the early mornings at around 7 a.m. to 8 a.m.; the cycle is very consistent with that in the similar study of Garland and Derwent [55]. This daily maximum is related to the influence of photochemical reactions with air pollutants, whereas the daily minimum is related to both ozone sinking by reaction with NO_2 [56] and ground deposition [57]. The night decadal exceedances up to 80 h for rural sites and up to 8 h for urban sites (Figure 5) can be explained by meteorological factors, i.e., the formation of an inversion layer at night trapping the ozone formed during the day or the wind speed and direction either facilitating the transport of ozone precursors or dispersing NO_2 and impeding its reaction with ozone. The decreased night-time exceedances for urban sites compared with rural sites can be explained by the sink reaction with NO_2, since this pollutant is more readily available in an urban background compared with in a rural background.

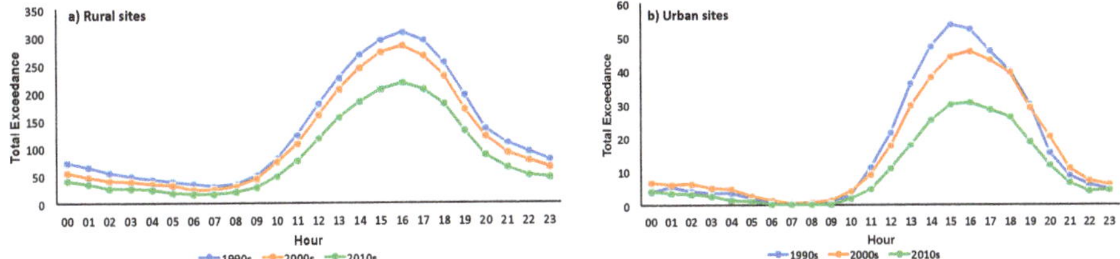

Figure 5. Hourly total ozone exceedances in (**a**) rural and (**b**) urban sites averaged for the decades of the 1990s, 2000s and 2010s. Total exceedance is calculated as the total number of hours at an ozone concentration ≥ 50 ppbv for each hour in a day, yearly averaged for each decade.

3.5. Weekly Variations of Ozone Exceedances over the Three Decades

Ozone exceedances for urban and rural sites follow a weekly pattern (Figure 6) since ozone formation depends on the temporal variations in precursor emissions. The trend is much the same as in previous studies [15,23], with the highest exceedances on weekends (e.g., on Saturday and Sunday) throughout the three decades. For the rural sites, there was a less marked variation in the exceedances, which remained constant throughout the whole week. However, for urban sites, the weekly variation is significant because of the strong variability in the emissions of precursors between the weekend and weekdays. The lower weekend NO_X concentrations due to lower traffic emissions reduce ozone scavenging, resulting in significantly higher weekend ozone concentrations than those on weekdays for urban sites [58].

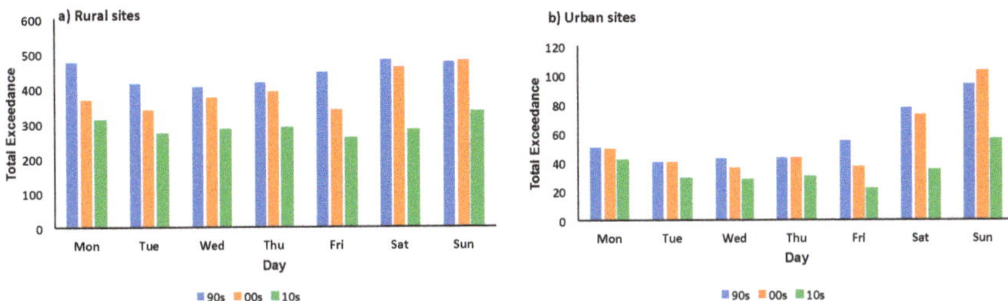

Figure 6. Daily total ozone exceedances in (**a**) rural and (**b**) urban sites averaged for the decades of the 1990s, 2000s and 2010s. Total exceedance is calculated as the total number of hours at an ozone concentration ≥ 50 ppbv for each day in a week, yearly averaged for each decade.

3.6. Frequency and Magnitude of Ozone Episodes

In order to determine the magnitude and frequency of the most severe ozone episodes in rural and urban sites, data were analysed for the years 1992, 1995, 1999, 2003, 2006, 2008, 2018 and 2019, years in which the numbers of exceedances were extremely high compared with those in other years. Exceedances were then considered for each summer month to determine the number of consecutive days ozone concentrations reached the limit of 50 ppbv.

The most severe ozone episodes in the rural and urban sites data (Table 2) show that the ozone concentrations with the greatest exceedances in May decreased over the years. For the urban sites, the number of consecutive days when exceedances were achieved were much lower compared with those for the rural sites. It is difficult to determine a single month and duration of episodes in this type of site due to the variable number of exceedances from one site to the other. After comparing the days in which each site reached a maximum ozone concentration for the longest period, it is deduced that three severe ozone episodes occurred in 2003, 2006 and 2019. This is because the highest concentrations reached in these years were far greater than those of previous years in every single site regardless of type. These events, defined as Case Studies 1–3, were evaluated for each site to assess their origin and magnitude (Table 3).

Table 2. Duration and magnitude of most severe episodes in rural and urban sites.

Year	Month	Duration (days)	Highest Mixing Ratios (ppb)	Site Type	Location
1992	May and June	16 and 9 + 9 [a]	125 and 125	Rural	Ha and Si
1995	August	17	133	Rural	YW
1995	May and August	6 and 3 + 4 [d]	104 and 102	Urban	LoB and CaC
1999	July	4	94	Urban	CaC
2003	August	8	108	Rural	Ha and LH
2003	August	7	116	Urban	SC
2006	July	4 + 7 [b]	114	Rural	LH
2006	July	4	101	Urban	CaC
2008	May	26 or 8 + 9 [c]	89	Rural	AH and YW
2008	May	27 or 9 + 10 [e]	90	Urban	CaC
2018	May	11	89	Rural	YW
2018	May	8	89	Urban	SC
2019	April	13	87	Rural	HM
2019	April	5	79	Urban	LeC

Note: [a] Two episodes in June lasting for 9 days each; [b] Two episodes in July lasting for 4 days and 7 days, respectively; [c] In some sites, a single long episode lasting for 26 days and in all other sites, two episodes lasting for 8 days and 9 days, respectively; [d] Two episodes in August lasting for 3 days and 4 days, respectively; [e] In some sites, a single long episode lasting for 27 days and in all other sites, two episodes lasting for 9 days and 10 days, respectively.

Case Study 1

The earliest weeks of August 2003 were particularly problematic in several European countries. A heat wave caused temperature records to be broken in France, Germany, Italy, Spain, the Netherlands and the UK. During August 9th and 10th, both England and Scotland broke their previous temperature records: 38.3 °C in England and 32.9 °C in Scotland [36]. This hot spell was characterized by anticyclonic conditions, where a high-pressure system moving around western Europe brought a hot, dry tropical continental air mass to the UK. This pattern occurred for much of the rest of the month [59]. Moderate to high concentrations of ozone were reported in most of the sites evaluated in this study. The highest concentrations were reported in Harwell (Ha), Lullington Heath (LH) and Southampton Centre (SC), with mixing ratios of over 103 ppbv reached in the afternoon of August 9th on each site. The period for this episode spans 8 consecutive days on average, starting between the 3rd and 4th of August.

Table 3. Magnitude and duration of case studies.

Sites	Case Study 1 August 2003		Case Study 2 July 2006		Case Study 3 April 2019	
	Days	O_3 (ppbv)	Days	O_3 (ppbv)	Days	O_3 (ppbv)
AH	9	79.5	6	112	10	77.4
Bo	11	99.9	11	95.8	N/A	N/A
Bu	5	66.3	3	<50	6	74.1
Ek	5	74.4	5 + 5	89.7	16	78.2
Gz	6	88.7	5 + 7	95.8	6	73.5
Ha	14	108.0	5 + 8	105.0	N/A	N/A
HM	7	68.3	6 + 7	71.3	16	86.3
Lb	11	88.7	5 + 6	70.3	7	81.5
LN	5	55.0	2	86.6	16	78.2
LH	10	108.0	5 + 8	114.2	14	76.0
Si	11	79.5	6 + 14	89.7	5	77.9
SV	4	75.4	2 + 3	92.8	8	75.7
YW	10	89.7	2 + 7	99.9	8	70.4
Be	4	65.2	3	95.8	5	76.5
Bi	1	68.3	6	85.6	N/A	N/A
CaC	9	85.6	5	100.9	4	72.9
LeC	8	70.3	4	78.5	8	78.9
LoB	6	90.7	5	90.7	4	61.9
SC	10	116.2	3	76.4	3	54.6

Note: N/A—no available data.

Case Study 2

Another record-breaking heat wave struck the UK during the summer of 2006. It was found that July 2006 was the warmest month on record in the UK [60], which was characterized by warm, sunny days associated with high pressure systems over northern Europe. This month was at least 1 degree Celsius hotter than in the previous case study, and during the first four days of the month, temperatures exceeded 30 °C across England and Wales, with the next few days back to more normal conditions. On July 11th, temperatures increased again, and an anticyclonic weather system became established over the UK. This middle part of the month was at its warmest and sunniest [60]. During this period, the ozone mixing ratios were even higher than in the previous case study, with the highest mixing ratios recorded in Lullington Heath (LH) at 114.2 ppbv and very high concentrations recorded in Aston Hill (AH), Harwell (Ha) and Cardiff Centre (CaC), all above 100 ppbv. The analysis of Case Studies 1 and 2 suggests that ozone formation and accumulation are directly dependent on the weather conditions.

Case Study 3

This episode was not characterized by a heat wave, in contrast to the previous two case studies. Even though it was described as the hottest Easter Weekend in the UK according to the BBC [61], the temperatures hardly reached 25 °C all over the Kingdom, and by the 26th of April, storm Hannah had already reached the UK and dissipated the bad weather conditions with winds of over 70 mph [62]. Moderate mixing ratios above 70 ppbv were recorded in all sites except for SC. The highest mixing ratio achieved was 86.3 ppbv in High Muffles (HM). It is evident that this case study also had the lowest recorded mixing ratios out of the three. However, the duration of this episode was much longer, lasting an average of 13 consecutive days on rural sites and 5 days on the urban ones.

We used the back-trajectories to determine from where the ozone precursors were coming and how they were affecting the sites in Case Study 3. The trajectories were followed by an incoming air parcel on a 96 h span towards three selected sites—Eskdatemuir (Ek), Lough Navar (LN) and Aston Hill (AH)—from April 17th to April 24th, arriving at 4 p.m. (local time), when the highest ozone mixing ratios were recorded for these sites. The sites chosen for this analysis were selected due to their location; an air parcel arriving in LN and AH (western sites) must travel through the south of the UK and possibly pass through other sites in this area. The same applies to Ek, since it is in the north of England. The trajectories for Ek and AH seem to have the same behaviour throughout the entire duration of the episode, passing through east Europe, the north of Germany and finally arriving in England, but on the 24th, the trajectory arriving in AH shifts suddenly from Europe to the Atlantic Ocean (see Figure S4). On the other hand, the trajectories arriving in LN change from day to day, coming in from France at the beginning of the episode, looping to the Atlantic Ocean on the 21st and then shifting again and arriving from central Europe on the day the highest ozone mixing ratios were reached and for the rest of the episode (see Figure S4). The highest ozone mixing ratio was recorded in HM on the 22nd at 8 p.m. local time. This site is on the Ek trajectory every single day of the episode. On the days prior to the 22nd, the trajectory loops over east and central Europe before arriving in Ek from a south-easterly direction (Figure S4). On the 22nd, the trajectories from both AH and Ek loop on each other from east Europe and separate over north Germany before arrival in the UK (see Figure S4). This means that the incoming air parcel on this day is the same for both sites, and when it finally arrives in HM, it has been carrying ozone precursors from four different countries. Two days later, the Ek trajectory changed its origin from east Europe to central Europe (Figure S4). It is then evident that the reason why this episode had such an effect on rural sites is purely the origin and transportation of ozone precursor emissions and not exclusively the weather conditions.

4. Conclusions

In this study, data from the Department of Environment, Food and Rural Affairs (Defra) was analysed in order to discuss the behaviour of ozone concentrations in the UK over a period from 1992 to mid-2019. This was done to explore how the frequency and magnitude of ozone episodes has changed over the years as a response to environmental policies to reduce pollutant emissions. It was found that maximum annual-mean ozone exceedances have decreased over the last three decades regardless of the type of site. The ozone exceedances in rural sites were found to be higher than those in the urban sites due to the lack of NO_X emissions to scavenge ozone. Ozone episodes have been shown to take place exclusively in the late spring and summer, when the sunlight is constant and the weather conditions prohibit pollution from dissipating. The diurnal variation of ozone exceedances shows a maximum during the mid-afternoon for both rural and urban sites. Additionally, ozone episodes usually are more frequent on the weekends due to a reduced removal effect of NO_X (emissions are lower on the weekends). It was also found that the day (in which ozone episodes are more frequent) has shifted from decade to decade due to the decrease in NO_X levels. In recent years, these episodes have become less marked, with exceedances remaining constant throughout the whole week regardless of the type of site. The adopted emission controls in both the UK and Europe have so far been successful in decreasing pollutant emissions. Ozone trends during the 2010s decade have become less visible, but further analyses must be carried out periodically with different methodologies in order to adopt and implement the policies to control the trend, be able to achieve the ozone objective limit of 50 ppbv and grant a better quality of life for both people and ecosystems.

Supplementary Materials: The following are available online at http://www.mdpi.com/2073-4433/11/5/534/s1, Table S1: The annual average ozone maximum and the trend of maximum ozone and average ozone for the period of 1992–2019. Figure S1: Monthly ozone exceedances in rural sites for last three decades, (a) 1990s, (b) 2000s and (3) 2010s. Figure S2: Monthly ozone exceedances in urban sites for last three decades, (a) 1990s, (b) 2000s and (3)

2010s. Figure S3: Trajectories arriving in Birmingham Centre (Bi), Cardiff Centre (CaC) and Leeds Centre (LeC) at 4 pm (local time) on 11 May 2008. Figure S4: Trajectories arriving in Eskdatemuir (Ek), Aston Hill (AH) and Lough Navar (LN) at 4 pm (local time) for the period of 17 April to 24 April 2019.

Author Contributions: F.M.R.D. and M.A.H.K. analyzed the data and wrote the paper; U.V. and D.E.S. conceived and designed the project; B.M.A.S. and E.D.G.S. analyzed the data. All authors have read and agreed to the published version of the manuscript.

Funding: DES and MAHK thank Natural Environment Research Council (NERC), Bristol ChemLabS and Primary Science Teaching Trust under whose auspices various aspects of this work were funded.

Acknowledgments: We thank the Department for Environment, Food and Rural Affairs (Defra) for supporting UK monitoring network data and National Oceanic and Atmospheric Administration (NOAA) on-line trajectory service for providing meteorological back trajectories.

Conflicts of Interest: The authors declare no conflict of interest.

References

1. Manahan, S.E. *Environmental Chemistry*, 8th ed.; CRC Press: Boca Raton, FL, USA, 2005; p. 244.
2. IPCC (Intergovernmental Panel on Climate Change). Working Group I Contribution to the IPCC Fifth Assessment Report "Climate Change 2013: The Physical Science Basis", Final Draft Underlying Scientific-Technical Assessment. 2013. Available online: http://www.ipcc.ch (accessed on 18 May 2020).
3. Dingenen, R.V.; Dentener, F.J.; Raes, F.; Krol, M.C.; Emberson, L.; Cofala, J. The global impact of ozone on agricultural crop yields under current and future air quality legislation. *Atmos. Environ.* **2009**, *43*, 604–618. [CrossRef]
4. Felzer, B.S.; Cronin, T.; Reilly, J.M.; Melillo, J.M.; Wang, X. Impacts of ozone on trees and crops. *Comptes Rendus Geosci.* **2007**, *339*, 784–798. [CrossRef]
5. Screpanti, A.; De Marco, A. Corrosion on cultural heritage buildings in Italy: A role for ozone? *Environ. Pollut.* **2009**, *157*, 1513–1520. [CrossRef] [PubMed]
6. Zhang, J.; Wei, Y.; Fang, Z. Ozone pollution: A major health hazard worldwide. *Front. Immunol.* **2019**, *10*, 2518. [CrossRef] [PubMed]
7. Heusser, K.; Tank, J.; Holz, O.; May, M.; Brinkmann, J.; Engeli, S.; Diedrich, A.; Framke, T.; Koch, A.; Groβhennig, A.; et al. Ultrafine particles and ozone perturb norepinephrine clearance rather than centrally generated sympathetic activity in humans. *Sci. Rep.* **2019**, *9*, 3641. [CrossRef]
8. Ebi, K.L.; McGregor, G. Climate Change, Tropospheric Ozone and Particulate Matter, and Health Impacts. *Environ. Health Perspect.* **2008**, *116*, 1449–1455. [CrossRef]
9. Lee, S.D.; Wolters, G.J.R.; Grant, L.D.; Schneider, T. *Atmospheric Ozone Research and Its Policy Implications*; Elsevier: Amsterdam, The Netherlands, 1989; p. 13.
10. Mazzuca, G.M.; Ren, X.; Loughner, C.P.; Estes, M.; Crawford, J.H.; Pickering, K.E.; Weinheimer, A.J.; Dickerson, R.R. Ozone production and its sensitivity to NOx and VOCs: Results from the DISCOVER-AQ field experiment, Houston 2013. *Atmos. Chem. Phys.* **2016**, *16*, 14463–14474. [CrossRef]
11. Clapp, L.J.; Jenkin, M.E. Analysis of the relationship between ambient levels of O_3, NO_2 and NO as a function of NOx in the UK. *Atmos. Environ.* **2001**, *35*, 6391–6405. [CrossRef]
12. Paoletti, E.; De Marco, A.; Beddows, D.C.S.; Harrison, R.M.; Manning, W.J. Ozone levels in European and USA cities are increasing more than at rural sites, while peak values are decreasing. *Environ. Pollut.* **2014**, *192*, 295–299. [CrossRef]
13. Monks, P.S.; Archibald, A.T.; Colette, A.; Cooper, O.; Coyle, M.; Derwent, R.; Fowler, D.; Granier, C.; Law, K.S.; Mills, G.E.; et al. Tropospheric ozone and its precursors from the urban to the global scale from air quality to short-lived climate forcer. *Atmos. Chem. Phys.* **2015**, *15*, 8889–8973. [CrossRef]
14. Sicard, P.; Serra, R.; Rossello, P. Spatiotemporal trends in ground-level ozone concentrations and metrics in France over the time period 1999–2012. *Environ. Res.* **2016**, *149*, 122–144. [CrossRef] [PubMed]

15. Khan, M.A.H.; Morris, W.C.; Galloway, M.; Shallcross, B.M.; Percival, C.J.; Shallcross, D.E. An estimation of the levels of stabilized Criegee intermediates in the UK urban and rural atmosphere using the steady-state approximation and the potential effects of these intermediates on tropospheric oxidation cycles. *Int. J. Chem. Kinet.* **2017**, *49*, 611–621. [CrossRef] [PubMed]
16. Jenkin, M. Trends in ozone concentration distributions in the UK since 1990: Local, regional and global influences. *Atmos. Environ.* **2008**, *42*, 5434–5445. [CrossRef]
17. Gaudel, A.; Cooper, O.R.; Ancellet, G.; Barret, B.; Boynard, A.; Burrows, J.P.; Clerbaux, C.; Coheur, P.-F.; Cuesta, J.; Cuevas, E.; et al. Tropospheric Ozone Assessment Report: Present-day distribution and trends of tropospheric ozone relevant to climate and global atmospheric chemistry model evaluation. *Elem. Sci. Anth.* **2018**, *6*, 39. [CrossRef]
18. Directive 2008/50/EC of the European Parliament and of the Council of 21 May 2008 on Ambient Air Quality and Cleaner Air for Europe. Available online: http://data.europa.eu/eli/dir/2008/50/2015-09-18 (accessed on 11 March 2020).
19. Air Quality Standards. European Commission—Environment. Available online: https://ec.europa.eu/environment/air/quality/standards.htm (accessed on 11 March 2020).
20. Department for Environment Food & Rural Affairs. The Air Quality Strategy for England, Scotland, Wales and Northern Ireland. 2007. Available online: http://www.defra.gov.uk/environment/airquality/strategy/pdf/air-qualitystrategy-vol1.pdf (accessed on 11 March 2020).
21. Jenkin, M.; Clemitshaw, K.C. Ozone and other secondary photochemical pollutants: Chemical processes governing their formation in the planetary boundary layer. *Atmos. Environ.* **2000**, *34*, 2499–2527. [CrossRef]
22. Derwent, R.G.; Simmonds, P.G.; Manning, A.J.; Spain, T.G. Trends over a 20-year period from 1987 to 2007 in surface ozone at the atmospheric research station, Mace Head, Ireland. *Atmos. Environ.* **2007**, *41*, 9091–9098. [CrossRef]
23. Jenkin, M.; Davies, T.J.; Stedman, J.R. The origin and day-of-week dependence of photochemical ozone episodes in the UK. *Atmos. Environ.* **2002**, *36*, 999–1012. [CrossRef]
24. Derwent, R.G.; Manning, A.J.; Simmonds, P.G.; Spain, T.G.; O'Doherty, S. Long-term trens in ozone in baseline and European regionally-polluted air at Mace Head, Ireland over a 30-year period. *Atmos. Environ.* **2018**, *179*, 279–287. [CrossRef]
25. National Air Quality Data Archive. Department for Environment Food & Rural Affairs. UK Air Information Resource. Available online: https://uk-air.defra.gov.uk/data/data_selector_service#mid (accessed on 11 March 2020).
26. EU Standard Methods for Monitoring and UK Approach, Air Quality Monitoring Methods, Department for Environment Food & Rural Affairs. UK Air Information Resource. Available online: https://uk-air.defra.gov.uk/networks/monitoring-methods?view=eu-standards (accessed on 11 March 2020).
27. The Air Quality Data Validation and Ratification Process. Department for Environment Food & Rural Affairs. Available online: https://uk-air.defra.gov.uk/assets/documents/Data_Validation_and_Ratification_Process_Apr_2017.pdf (accessed on 19 December 2019).
28. QA/QC Procedures for the UK Automatic Urban and Rural Air Quality Monitoring Network (AURN) Report to Defra and the Developed Administrations (2009) (AEAT/ENV/R/2837). Available online: https://uk-air.defra.gov.uk/assets/documents/reports/cat13/0910081142_AURN_QA_QC_Manual_Sep_09_FINAL.pdf (accessed on 11 March 2020).
29. Draxler, R.R.; Rolph, G.D. HYSPLIT (Hybrid Single-Particle Lagrangian Integrated Trajectory) NOAA Air Resourcses Laboratory. Available online: http://www.arl.noaa.gov/ready/hysplit4.html (accessed on 18 May 2020).
30. Air Quality Expert Group. Ozone in the United Kingdom. Department for the Environment, Food and Rural Affairs, 2009. Available online: https://uk-air.defra.gov.uk/assets/documents/reports/aqeg/aqeg-ozone-report.pdf (accessed on 18 May 2020).
31. Tørseth, K.; Aas, W.; Breivik, K.; Fjaeraa, A.M.; Fiebig, M.; Hjellbrekke, A.G.; Lund Myhre, C.; Solberg, S.; Yttri, K.E. Introduction to the European Monitoring and Evaluation Programme (EMEP) and observed atmospheric composition change during 1972–2009. *Atmos. Chem. Phys.* **2012**, *12*, 5447–5481. [CrossRef]
32. Simpson, D.; Arneth, A.; Mills, G.; Solberg, S.; Uddling, J. Ozone-the persistent menace: Interactions with the N cycle and climate change. *Curr. Opin. Environ. Sustain.* **2014**, *9–10*, 9–19. [CrossRef]
33. Li, Y.; Lau, A.K.H.; Fung, J.C.H.; Zheng, J.; Liu, S. Importance of NO_x control for peak ozone reduction in the Pearl River Delta region. *J. Geophys. Res. Atmos.* **2013**, *118*, 9428–9443. [CrossRef]

34. Yan, Y.; Pozzer, A.; Ojha, N.; Lin, J.; Lelieveld, J. Analysis of European ozone trends in the period 1995–2014. *Atmos. Chem. Phys.* **2018**, *18*, 5589–5605. [CrossRef]
35. Solberg, S.; Hov, Ø.; Søvde, A.; Isaksen, I.S.A.; Coddeville, P.; De Backer, H.; Forster, C.; Orsolini, Y.; Uhse, K. European surface ozone in the extreme summer 2003. *J. Geophys. Res. Atmos.* **2008**, *113*, D07307. [CrossRef]
36. Burt, S. The August 2003 heatwave in the United Kingdom Part 1—Maximum temperatures and historical precedents. *Weather* **2004**, *59*, 199–208. [CrossRef]
37. Kendon, M. Has there been a recent increase in UK weather records? *Weather* **2014**, *69*, 327–332. [CrossRef]
38. Nightingale, B.; Allsopp, K. Invasion of Red-footed Falcons in spring 1992. *Brit. Birds* **1994**, *87*, 223–231.
39. Hulme, M. The climate in the UK from November 1994 to October 1995. *Weather* **1997**, *52*, 242–257. [CrossRef]
40. McCarthy, M.; Christidis, N.; Dunstone, N.; Fereday, D.; Kay, G.; Klein-Tank, A.; Lowe, J.; Petch, J.; Scaife, A.; Stott, P. Drivers of the UK summer heatwave of 2018. *Weather* **2019**, *74*, 390–396. [CrossRef]
41. Monks. P.S. A review of the observations and origins of the spring ozone maximum. *Atmos. Environ.* **2000**, *34*, 3545–3561. [CrossRef]
42. Pope, R.J.; Butt, E.W.; Chipperfield, M.P.; Doherty, R.M.; Fenech, S.; Schmidt, A.; Arnold, S.R.; Savage, N.H. The impact of synoptic weather on UK surface ozone and implications for premature mortality. *Environ. Res. Lett.* **2016**, *11*, 124004. [CrossRef]
43. UNECE. The 1999 Gothenburg Protocol to Abate Acidification, Eutrophication and Ground-Level Ozone. Available online: www.unece.org/env/lrtap/multi_h1.htm (accessed on 11 March 2020).
44. Hanna, E.; Mayes, J.; Beswick, M.; Prior, J.; Wood, L. An analysis of the extreme rainfall in Yorkshire, June 2007, and its rarity. *Weather* **2008**, *63*, 253–260. [CrossRef]
45. British Isles Weather, June 2011. University of Reading, Department of Meteorology. Available online: http://www.met.rdg.ac.uk/~brugge/diary2011.html#201106 (accessed on 12 March 2020).
46. Jaroszweski, D.; Hooper, E.; Baker, C.; Chapman, L.; Quinn, A. The impacts of the 28 June 2012 storms on UK road and rail transport. *Meteorol. Appl.* **2015**, *22*, 470–476. [CrossRef]
47. Parry, S.; Barker, L.; McKenzie, A.; Clemas, S. Hydrological Summary for the United Kingdom June 2013. NERC/Centre for Ecology and Hydrology. Available online: http://nora.nerc.ac.uk/id/eprint/502622/1/HS201306.pdf (accessed on 12 March 2020).
48. Parry, S.; Barker, L.; McKenzie, A.; Clemas, S. Hydrological summary for the United Kingdom June 2014. NERC/Centre for Ecology and Hydrology. Available online: http://nora.nerc.ac.uk/id/eprint/507823/1/HS_201406.pdf (accessed on 12 March 2020).
49. Kendon, M.; McCarthy, M.; Jevrejeva, S.; Matthews, A.; Legg, T. State of the UK Climate 2015. Met Office: Exeter, UK. Available online: https://www.metoffice.gov.uk/binaries/content/assets/metofficegovuk/pdf/weather/learn-about/uk-past-events/state-of-uk-climate/mo-state-of-uk-climate-2015-v3.pdf (accessed on 12 March 2020).
50. Kendon, M.; McCarthy, M.; Jevrejeva, S.; Matthews, A.; Legg, T. State of the UK Climate 2016. Met Office: Exeter, UK. Available online: https://www.metoffice.gov.uk/binaries/content/assets/metofficegovuk/pdf/weather/learn-about/uk-past-events/state-of-uk-climate/mo-state-of-uk-climate-2016-v4.pdf (accessed on 12 March 2020).
51. Kendon, M.; McCarthy, M.; Jevrejeva, S.; Matthews, A.; Legg, T. State of the UK climate 2017. *Int. J. Clim.* **2018**, *38*, 1–35. [CrossRef]
52. Kendon, M.; McCarthy, M.; Jevrejeva, S.; Matthews, A.; Legg, T. State of the UK climate 2018. *Int. J. Clim.* **2019**, *39*, 1–55. [CrossRef]
53. Wet Weather June 2019. Met Office. Available online: https://www.metoffice.gov.uk/binaries/content/assets/metofficegovuk/pdf/weather/learn-about/uk-past-events/interesting/2019/2019_006_rainfall_lincolnshire.pdf (accessed on 19 December 2019).
54. Shukla, J.B.; Misra, A.K.; Sundar, S.; Naresh, R. Effect of rain on removal of a gaseous pollutant and two different particulate matters from the atmosphere of a city. *Math. Comp. Model.* **2008**, *48*, 832–844. [CrossRef]
55. Garland, J.A. and Derwent, R.G. Destruction at the ground and the diurnal cycle of concentrations of ozone and other gases. *Q. J. R. Meteorol. Soc.* **1979**, *105*, 169–183. [CrossRef]
56. Sillman, S. The relation between ozone, NOx and hydrocarbons in urban and polluted rural environments. *Atmos. Environ.* **1999**, *33*, 1821–1845. [CrossRef]
57. Galbally, I.E.; Roy, C.R. Destruction of ozone at the Earth's surface. *Q. J. R. Meterol. Soc.* **1980**, *106*, 599–620. [CrossRef]

58. Henry, R.F. Weekday/weekend differences in gasoline related hydrocarbons at coastal PAMS sites due to recreational boating. *Atmos. Environ.* **2013**, *75*, 58–65. [CrossRef]
59. Kent, A. Air Pollution Forecasting: Ozone Pollution Episode Report (August 2003). Department for Environment Food & Rural Affairs, UK. Available online: https://uk-air.defra.gov.uk/assets/documents/reports/cat12/o3_episode_august2003.pdf (accessed on 18 May 2020).
60. Prior, J.; Beswick, M. The record breaking heat and sunshine of July 2006. *Weather* **2007**, *62*, 174–182. [CrossRef]
61. UK Weather: Hottest Easter Monday on Record. *BBC News*, 22 April 2019. Available online: https://www.bbc.co.uk/news/uk-48013791 (accessed on 12 March 2020).
62. Storm Hannah 26 to 27 April 2019. Met Office. Available online: https://www.metoffice.gov.uk/binaries/content/assets/metofficegovuk/pdf/weather/learn-about/uk-past-events/interesting/2019/2019_005_storm_hannah.pdf (accessed on 12 March 2020).

Article

A Stratospheric Intrusion-Influenced Ozone Pollution Episode Associated with an Intense Horizontal-Trough Event

Yiping Wang [1], Hongyue Wang [1] and Wuke Wang [2,*]

1 School of Atmospheric Sciences, Nanjing University, Nanjing 210023, China; wypfyyd@nju.edu.cn (Y.W.); mg1728018@smail.nju.edu.cn (H.W.)

2 Department of Atmospheric Science, China University of Geosciences, Wuhan 430074, China

* Correspondence: wangwuke@cug.edu.cn

Received: 15 December 2019; Accepted: 28 January 2020; Published: 4 February 2020

Abstract: Ozone pollution is currently a serious issue in China. As an important source of tropospheric ozone, the stratospheric ozone has received less concern. This study uses a combination of ground-based ozone measurements, the latest ERA5 reanalysis data as well as chemistry-climate model and Lagrangian Particle Dispersion Modeling (LPDM) simulations to investigate the potential impacts of stratospheric intrusion (SI) on surface ozone pollution episodes in eastern China. Station-based observations indicate that severe ozone pollution occurred from 27 April to 28 April 2018 in eastern China, with maximal values over 140 ppbv. ERA5 meteorological and ozone data suggest that a strong horizontal-trough exists at the same time, which leads to an evident SI event and brings ozone-rich air from the stratosphere to the troposphere. Using a stratospheric ozone tracer defined by NCAR's Community Atmosphere Model with Chemistry (CAM-Chem), we conclude that this SI event contributed about 15 ppbv (15%) to the surface ozone pollution episode during 27–28 April in eastern China. The potential impacts of SI events on surface ozone variations should be therefore considered in ozone forecast and control.

Keywords: tropospheric ozone; stratospheric intrusion; horizontal-trough

---

1. Introduction

Tropospheric ozone is an important greenhouse gas contributing to global warming [1,2] and also a main pollutant harmful to human health and crop productivity [3,4]. A significant decrease of ozone concentrations has been observed in the eastern United States and parts of Europe due to precursor emission controls [2,5,6]. At the same time, although the Chinese government has strictly controlled industrial emissions since 2011, ozone concentrations in eastern China have increased over the past years [7,8]. Ozone pollution episodes have been reported frequently in eastern China, and ozone pollution is now a serious issue in China [6,7,9]. Surface ozone is related to complex photochemical reactions involving anthropogenic and biogenic emitted volatile organic compounds (VOCs) and nitrogen oxides (NOx) [10–12]. While the anthropogenic emitted NOx decreased significantly due to government control and the VOCs emissions changed little [7,8,12], the exact reason for elevated ozone and frequently occurring ozone pollution episodes in eastern China are still not clear.

Besides chemical production, tropospheric ozone can also be transported from the stratosphere through the process of stratospheric intrusion (SI) [13–17]. As another important source of tropospheric ozone, SI contributes 20%–30% to the tropospheric ozone budget in the mid-latitudes of the Northern Hemisphere [14,18,19]. In cases of intense SIs, the stratospheric ozone-rich air may be transported rapidly from the lower stratosphere to the lower troposphere and near the surface [20–23]. Such deep SI events may cause a steep ozone increase near the surface and lead to ground-level ozone pollution [24–28]. Whether such deep SI events contribute to the recently reported ozone pollution episodes in eastern China is an important question waiting to be answered. Quantifying the relative contribution of SI events to the elevated ozone is important for surface ozone forecasting and control in eastern China.

In the extratropics, SIs always form near the vicinity of extratropical cyclones or baroclinic eddies, through processes like tropopause folding, wave-breaking or cut-off lows [13,29,30]. SIs are often associated with westerly jet stream or frontal activities [17,31,32]. Most of these SI processes occur in areas with high latitudes or high altitudes [33–35]. In East Asia, studies of STE mainly focus on the Qinghai-Tibet Plateau and northeast China [15,36–39]. The potential effects of SI events on surface ozone in eastern China, however, has been less investigated. A recent study found that SI events occur frequently during the summer and contribute about 10 ppbv to surface ozone in eastern China [40]. This poses a question about the potential impacts of SI events on surface ozone variations in eastern China, whereas human activities are particularly active and heavy air pollution happens frequently [6,9,41].

This study investigates a severe ozone pollution episode that occurred during 28–29 April 2018 using the station-based surface ozone measurements as well as the latest ERA5 reanalysis ozone data. In particular, the potential contribution of SIs to surface ozone in the spring season (MAM) is estimated with the aid of a stratospheric ozone tracer of the CAM-Chem model. The transport mechanism is also analyzed using the ERA5 meteorological reanalysis data.

2. Data and Model Description

2.1. Observational Data Sets

Hourly surface ozone and CO concentrations for the year 2018 were obtained from the public website of the China Ministry of Ecology and Environment: beijingair.sinaapp.com/. The network has had 1500 monitoring stations since 2017 including about 330 cities.

2.2. ERA5 Reanalysis

ERA5 (the fifth generation of ECMWF atmospheric reanalysis) [42] ozone and meteorological data are used in this study for analyzing the ozone and meteorological conditions during the SI events. ERA5 was produced using 4D-Var data assimilation in CY41R2 of ECMWF's Integrated Forecast System (IFS), with 137 hybrid sigma/pressure (model) levels from the surface to 0.01 hPa. The horizontal resolution of ERA5 for the high resolution realisation (HRES) is about 31 km, 0.28125 degrees. More details of the ERA5 data can be found at its website https://confluence.ecmwf.int/display/CKB/ERA5%3A+data+documentation. Within the framework of the Copernicus Atmosphere Monitoring Service (CAMS) atmospheric composition forecast, IFS was found capable of predicting ozone increases in the troposphere during deep SI events [17]. ERA5 assimilated many ozone observations (satellite and in-situ) including MLS (Microwave Limb Sounder), OMI (Ozone Monitoring Instrument) ozone data, which is of great importance to the determination of STE (Stratosphere Troposphere Exchange). ERA5 ozone shows good agreement with in-situ and satellite measurements in the upper troposphere and lower stratosphere [40]. However, its accuracy in the lower troposphere is relatively poor [40]. That is because the chemical scheme of ozone in ERA5 is completely dependent on the parameterization of ozone source/sink (https://www.ecmwf.int/en/elibrary/16648-part-iv-physical-processes). In addition, due to the lack of in-situ and satellite observations, the tropospheric ozone in ERA5 is not well constrained.

2.3. CAM-Chem Simulations

The Community Atmosphere Model with Chemistry (CAM-Chem) simulation used in this study is performed by NCAR , which is publicly available at https://www.acom.ucar.edu/cam-chem/cam-chem.shtml. It is driven by specified dynamics, with meteorological fields from MERRA2 reanalysis.

It is run at 0.9° × 1.25° horizontal resolution with 56 vertical levels. The chemistry mechanism used is the MOZART-T1. In particular, the O_3S variable, a stratospheric ozone tracer relative to the tropopause, is defined by the CAM-Chem model to estimate the stratospheric contribution to tropospheric ozone. O_3S is set to the ozone mixing ratio in the stratosphere and is destroyed below the tropopause at the same rate as ozone [43]. More details of the simulation are described on the website https://wiki.ucar.edu/display/camchem/CESM2.1%3ACAM-chem+as+Boundary+Conditions.

2.4. LPDM Simulations

LPDM (Lagrangian Particle Dispersion Modeling) is conducted using the HYSPLIT (Hybrid Single-Particle Lagrangian Integrated Trajectory) model [9,44]. In the model, the residence time and position of particles released at a receptor can be calculated according to the meteorological field. In this study, the model was run 24-h backwards with 3000 particles released at an altitude of 100 m above Hangzhou at 16:00 (local time) on 27 April 2018. The residence time of particles was used to identify the "footprint" retroplume. More details of the LPDM can be found in Ding et al. [45].

3. Results

3.1. Surface Ozone Pollution during 25–28 April 2018

Figure 1 shows the horizontal distribution of observed surface ozone concentration in eastern China (from the China Ministry of Ecology and Environment observations). High ozone concentrations can be seen in the cross-region between Jiangsu, Zhejiang and Anhui provinces. In particular, severe ozone pollution occurred on 27 April 2018 and 28 April 2018, with maximal ozone concentrations over 100 ppbv in the cross-region and over 140 ppbv in the Jiaxing City.

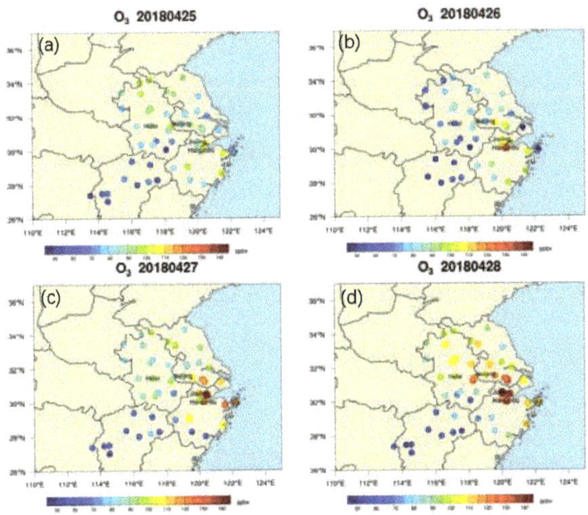

Figure 1. Spacial distribution of ground-level ozone concentration in eastern China, including Jiangsu, Zhejiang, Anhui, Jiangxi provinces and Shanghai, from 25 April 2018 to 28 April 2018.

The time evolution of surface ozone in the cities of Nanjing (the capital of Jiangsu Province), Hangzhou (the capital of Zhejiang province), Hefei (the capital of Anhui province) and Jiaxing (the city shows the highest ozone concentration) is shown in Figure 2. In Nanjing, ozone concentration exceeded 100 ppbv on 27–28 April 2018, with maximum values of 110 ppbv on 28 April 2018. In Hangzhou, ozone pollution (> 100 ppbv) occurred on 26–28 April 2018, with maximum values of 120 ppbv on 26 April 2018. Severe ozone pollution occurs from 25 April 2018 to 30 April 2018 in Jiaxing, with ozone concentration exceeds 140 ppbv on 27 and 28 April 2018. Hefei also shows high ozone concentrations above 100 ppbv on 25, 27 and 28 April 2018, with a maximum of 114 ppbv on 28 April 2018.

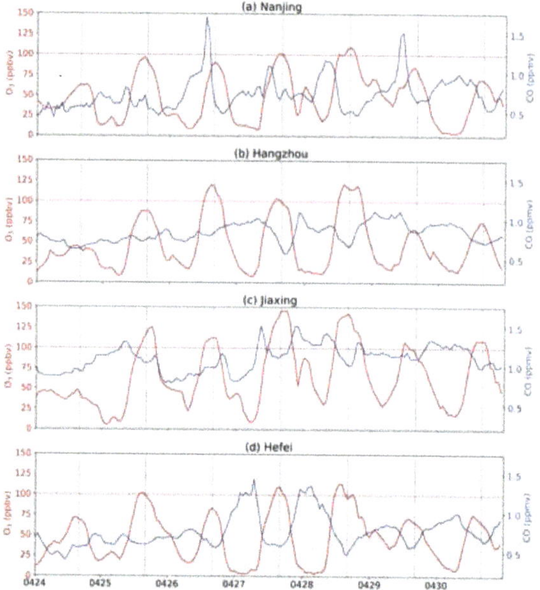

Figure 2. Time series of ground-level ozone (red lines) and CO (blue lines) concentration in eastern China from 24 April 2018 to 30 April 2018. (**a**) Nanjing, (**b**) Hangzhou, (**c**) Jiaxing and (**d**) Hefei. The horizontal red dashed line indicates the value of 100 ppbv of ozone concentration. The vertical black dotted lines indicate 16:00 of each day (local time), which is approximately the time of the maximal values of the diurnal cycle.

Figure 3 shows the mean diurnal cycle of ozone and CO averaged over the month (April 2018, solid lines) and of the severe ozone pollution days (27–28 April 2018, dashed lines) in the cities of Nanjing, Hangzhou, Hefei and Jiaxing. The maximal ozone concentration of the monthly mean diurnal cycle is no more than 75 ppbv in all four cities, while the ozone peak exceeded 100 ppbv on 27–28 April 2018. In normal conditions (mean of the month), ozone in selected cities show a clear diurnal cycle with a daily maximum of about 16:00 (local time) due to the photochemical production of ozone. On 27–28 April 2018, a delayed secondary peak of ozone is evident in all the four selected cities (Figures 2 and 3). This indicates an extra mechanism besides the normal photochemical production. At the same time, concentrations of CO are relatively low and anti-correlated with ozone on the ozone polluted days (27–28 April 2018). While low values of CO are a tracer for air of stratospheric origin, this suggests that the ozone pollution episode on 27–28 April 2018 might be related to stratospheric intrusion.

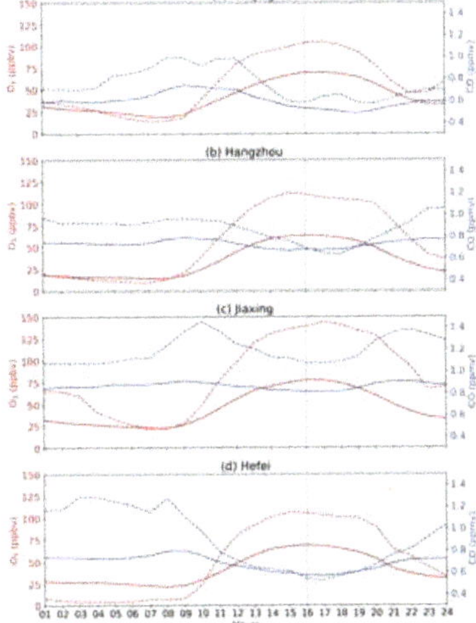

Figure 3. Mean diurnal cycle of ozone (red) and CO (blue) averaged over the month (April 2018, solid lines) and for the severe ozone pollution days (27–28 April 2018, dashed lines). (**a**) Nanjing, (**b**) Hangzhou, (**c**) Jiaxing and (**d**) Hefei. The vertical black dotted lines indicate 16:00 of local time, which is approximately the peak of the diurnal cycle.

3.2. Meteorological Conditions

The synoptic conditions over east Asia in the upper atmosphere (at 300 hPa) are shown in Figure 4. On 25 April 2018, two upper-level troughs are evident with one shallow trough over Mongolia and north China and the other deep trough over the region between northeastern China and Japan. The shallow trough moves faster than the deep one and catches up with the deeper trough. Subsequently, the two upper-level troughs combine together and formed an intense horizontal-trough on 26 April 2018. On 27 April 2018, the centre of the two upper-level trough merged completely and the connection between the centre and the western part of the horizontal-trough became weaker. Later, the centre of the deep trough moved eastward and the western part formed a cut-off low on 28 April 2018. The area of interest is on the south edge of the deep trough (see figure on 27 April 2018), which brings air from high latitudes.

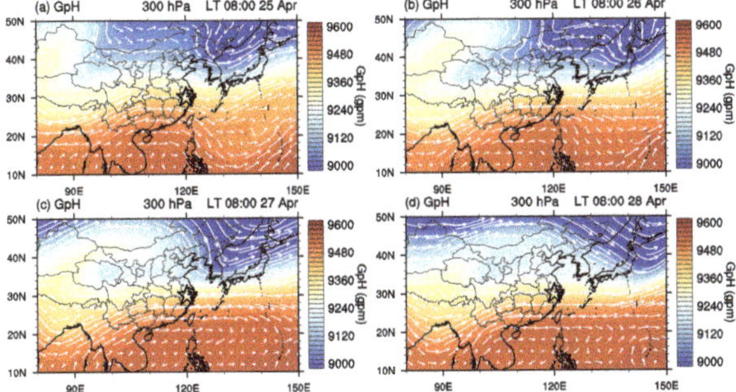

Figure 4. Spacial distribution of geopotential height (shading) and horizontal winds (arrows) at 300 hPa over east Asia from 25 April 2018 to 28 April 2018. The geopotential height and horizontal winds data are from ERA5. The black star in each panel indicates the area of interest in this study.

Figure 5 shows the synoptic conditions over east Asia at 500 hPa. Consistent with the upper-level conditions shown in Figure 4, two troughs can be seen at 500 hPa on 25 April 2018 with an earlier strong one over the region between northeastern China and Japan and a later weak one over Mongolia and north China. The later one moved faster and was enhanced during its moving. From 26 April to 27 April 2018, the two troughs merged together and formed a very deep trough. On 27 April 2018, the area of interest was right behind and influenced by the deep trough, which brought air from higher latitudes.

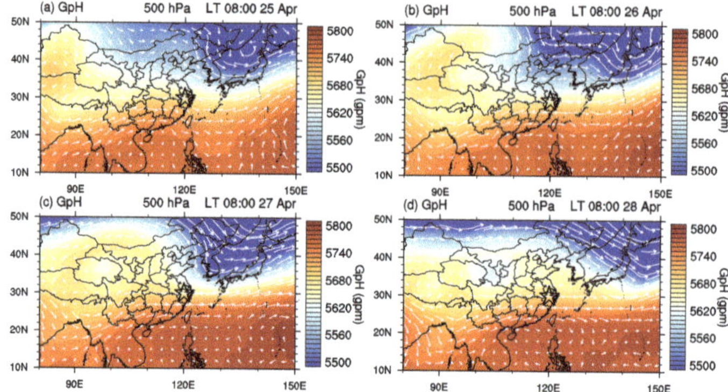

Figure 5. Spatial distribution of geopotential height (shading) and horizontal winds (arrows) at 500 hPa over east Asia from 25 April 2018 to 28 April 2018. The geopotential height and horizontal winds data are from ERA5. The black star in each panel indicates the area of interest of this study.

Figure 6 shows the spacial distribution of potential vorticity (PV) at 300 hPa. In the absence of friction and heat sources, PV is conserved along trajectories, which allows it to be used as a tracer in upper-level dynamics. In particular, PV is broadly used to separate the stratospheric and tropospheric air masses due to the significant difference of the static stability between them. Following previous studies [40], we used 2 PVU to mark the dynamical tropopause. On 25 April 2018, high PV values can be seen over Mongolia and north China as well as the region between northeastern China and Japan, associated with the two upper-level troughs as seen in Figure 3. Along with the horizontal-trough development, a broad quasi-horizontal band of high PV values can be seen from the west to the east of China on 26 and 27 April 2018. It is noteworthy that there are two belts in this high PV band, with one between 30° and 40° N and the other between 20° and 30° N. The later one crossed the area of interest in this study. This suggests a potential impact of stratospheric ozone intrusion on the tropospheric ozone over the area of interest.

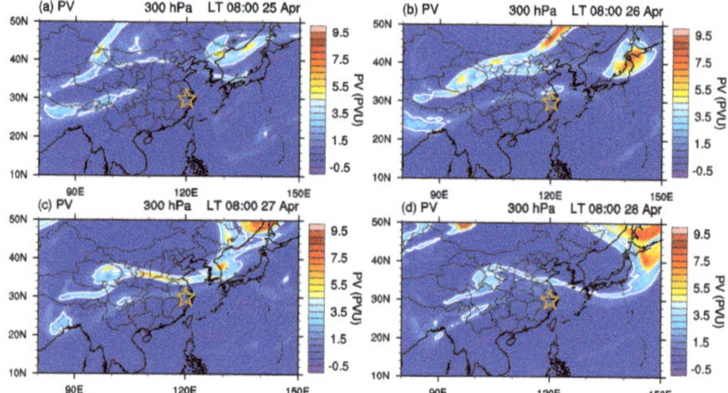

Figure 6. Spatial distribution of potential vorticity (PV) at 300 hPa over east Asia from 25 April 2018 to 28 April 2018. PV is calculated based on ERA5 data. The white contour line indicates the dynamical tropopause of 2 PVU (1 PVU = 10^{-6} m^2 s^{-1} K kg^{-1}). The red star in each panel indicates the area of interest in this study.

3.3. Stratospheric Ozone Intrusion to the Troposphere

To illustrate the downward transport, the ERA5 ozone data are used for analysing the ozone distribution in the upper troposphere and lower stratosphere (UTLS) region. Figure 7 shows the horizontal distribution of ozone at 300 hPa from 25 April 2018 to 28 April 2018. The spatial pattern of high ozone values is very similar to the PV pattern shown in Figure 5. This confirms the transport of ozone-rich air from the stratosphere to the troposphere associated with the upper-level trough development (Figure 3). The area of interest is on the edge of the high ozone concentration band on 25–26 April 2018 and near the centre of this high ozone concentration band on 27–28 April 2018.

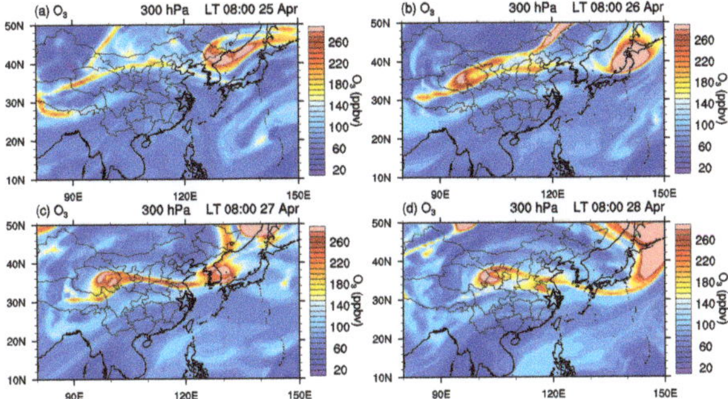

Figure 7. Spatial distribution of ERA5 ozone at 300 hPa over east Asia from 25 April 2018 to 28 April 2018. The black star in each panel indicates the area of interest in this study.

The downward transport of the stratospheric ozone-rich air is illustrated in Figure 8. On 25 April 2018, the stratospheric ozone-rich air penetrated to about 300 hPa at the longitude of 100° E. Later, the ozone-rich air appeared eastward and further down to the troposphere accompanying the eastward movement of the upper-level trough (Figure 3). On 27 April, the ozone-rich air reached the area of interest and extended to 400 hPa deeper in the troposphere. Ozone-rich air persisted above the area of interest at about 400 hPa on 28 April 2018.

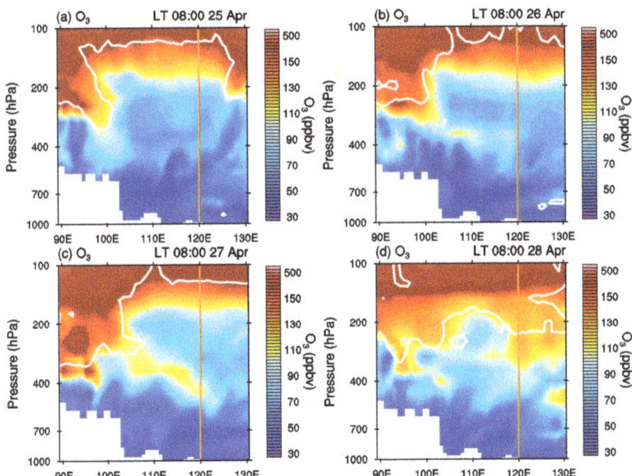

Figure 8. Longitude-height cross sections of ERA5 ozone at 30° N from 25 April 2018 to 28 April 2018. (**a**) 25 April, (**b**) 26 April, (**c**) 27 April and (**d**) 28 April 2018. The white line indicates the dynamical tropopause (2 PVU). The vertical line indicates the longitude (120° E) of Hangzhou.

The transport of ozone-rich air from the stratosphere to the troposphere is also confirmed by the O_3S tracer. Figure 9 shows the longitude-height cross-section of the O_3S/O_3 ratio from the CAM-Chem simulation. Consistent with ERA5 ozone shown in Figure 8, high values of O_3S/O_3 are seen from the stratosphere to about 400 hPa at the longitude of 100° E on 25 April 2018. As the up-level trough moves eastward, high values of O_3S/O_3 inject deeper in the troposphere and get to the area of interest on 27 April 2018. A centre of high O_3S/O_3 ratios is evident around 700 hPa over the longitude from 120° E to 130° E on 27 and 28 April 2018.

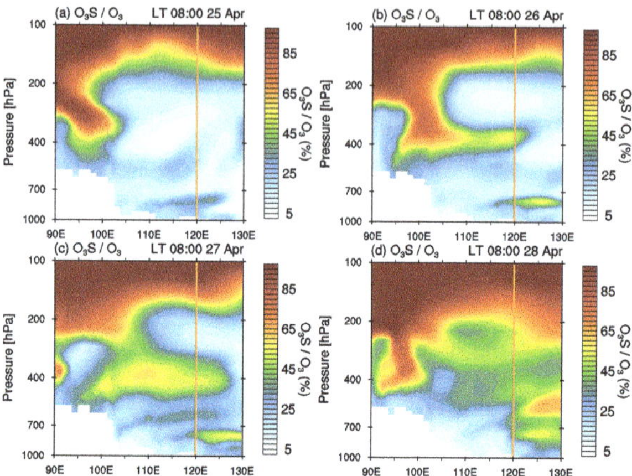

Figure 9. Longitude-height cross sections of O_3S/O_3 ratio at 30° N from CAM-Chem simulation for the period 25–28 April 2018. (**a**) 25 April, (**b**) 26 April, (**c**) 27 April and (**d**) 28 April 2018. The vertical line indicates the longitude (120° E) of the area of interest.

3.4. SI Impacts on Surface Ozone

So far, we find that the severe ozone pollution episode from 27 to 28 April 2018 might be related to an SI event. From the discussion above, the SI event is evident and the transport of ozone-rich air from the lower stratosphere to the troposphere is clear. The stratospheric ozone-rich air reached the middle troposphere below 400 hPa. However, whether such enhanced ozone concentrations in the troposphere can be mixed with the planetary boundary layer (PBL) and influence the surface is still not clear.

To illustrate the vertical mixing between the free troposphere and the PBL, the divergence of water vapour as well as the vertical velocity are shown in Figure 10. Positive values of water vapour divergence dominated from 26 to 27 April 2018. In particular, a strong divergence centre from 600 hPa down to the surface can be seen on 26 and 27 April 2018, which indicates the sink of the dry air. This suggests that the ozone-rich and dry air may move further down from the middle troposphere to the lower troposphere. On 28 April 2018, weak convergence occurred in the lower troposphere, which indicates a possible vertical mixing between the PBL and the upper levels.

Consistent with the water vapour divergence, downward motion of air masses is evident from 27 to 28 April 2018 as shown in the vertical distribution of vertical velocity. In particular, positive values of vertical velocity (indicating sink of air) can be seen from 400 hPa to the lower troposphere from 27 April to the morning of 28 April 2018. This strengthens the case that the ozone-rich and dry air is transported from the upper troposphere down to the PBL. On the afternoon of 28 April 2018, upward motion can be seen indicating unstable conditions and subsequent vertical mixing between the PBL and upper levels.

The vertical mixing between the free troposphere and the PBL can also be confirmed by the 'footprint' of air mass as shown by LPDM simulations. A 24-h backward integration of LPDM indicates that particles released at 16:00 (local time) in Hangzhou mainly originated from northern China (Figure 11a). Vertically, seen from the northwest-southeast transect of the retroplume (Figure 11b), the particles originated from up to 5 km (close to 500 hPa). Especially, the retroplume distributes vertically from the surface to about 700 hPa, which indicates strong vertical-mixing between the PBL and the free troposphere. The stratospheric ozone-rich air was therefore transported from the free troposphere to the surface.

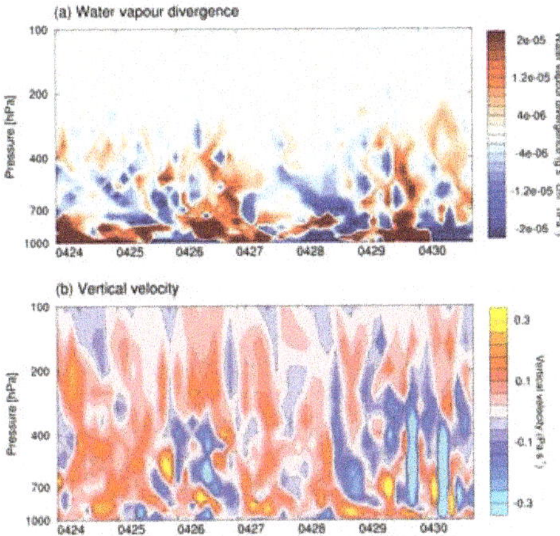

Figure 10. Time evolution of vertical distributions of water vapour divergence (**a**, g s^{-1} cm^{-2} hPa^{-1}) and vertical velocity (**b**, Pa/s), at Hangzhou (120° E, 30.2° N) from 24 April 2018 to 30 April 2018. The water vapour divergence and vertical velocity are calculated by ERA5 data.

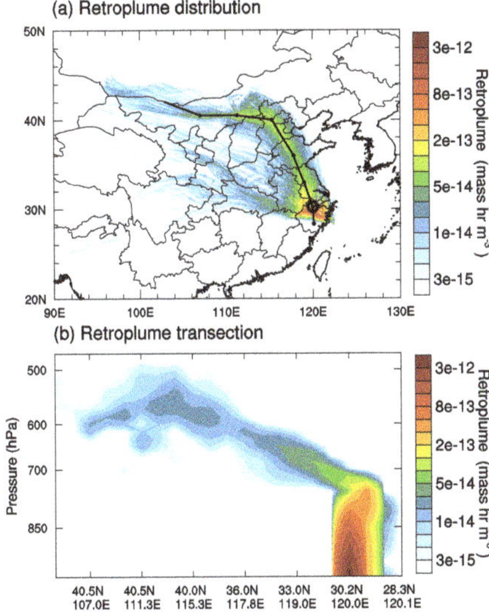

Figure 11. (**a**) 24-h backward column retroplume (integrated for all levels) of particles released at the altitude of 100 m above Hangzhou (120° E, 30.2° N) at 16:00 (local time) on 28 April 2018. (**b**) Vertical distribution of particle retroplume along the air path as indicated by the black dotted line in Figure 11a.

To quantify the potential contribution of SI events to surface ozone, Figure 12 shows the horizontal distribution of O_3S (the stratospheric ozone tracer) over eastern China on 27 and 28 April 2018, at 14:00 (local time). A centre of high O_3S values exists on 27 and 28 April 2018 in the area of interest. The SI event contributes about 15 ppbv to the surface ozone in the cross region between Jiangsu, Zhejiang and Anhui provinces. In particular, the centre with highest O_3S values of 17 ppbv reached Hangzhou on 28 April 2018. As reported by previous studies, O_3S gives an upper limit to the stratospheric ozone contribution [43]. The relative contribution of O_3S to the CAM-Chem simulated ozone concentration (O_3S/O_3) is also analysed. Figure 11a,b show the horizontal distribution of O_3S/O_3 ratio over eastern China on 27 and 28 April 2018, at 14:00 (local time). The O_3S/O_3 ratio is also relatively high in the area of interest. O_3S is over 15% of ozone in the area of interest on 27 and 28 April 2018.

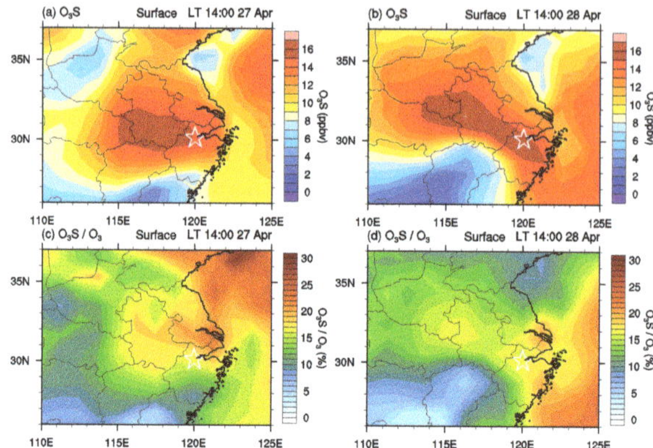

Figure 12. Spacial distribution of the CAM-Chem simulated O_3S (**a**,**b**) and O_3S/O_3 ratio (**c**,**d**) in eastern China at surface on 27 and 28 April 2018. The white star in each panel indicates the location of Hangzhou.

The time series of ozone concentration, as well as O_3S from the CAM-Chem simulation in Najing, Hangzhou, Jiaxing and Hefei from 22 to 30 April 2018, are presented in Figure 13. The CAM-Chem model simulates the ozone pollution episode relatively well, although the absolute value is lower than observations. According to the CAM-Chem simulation, the stratospheric ozone intrusion contributes about 15 ppbv (15%) to the ozone pollution episode from 27 to 28 April 2018. Such a significant contribution cannot be neglected and should be considered in surface ozone forecasting and developing emission control strategies.

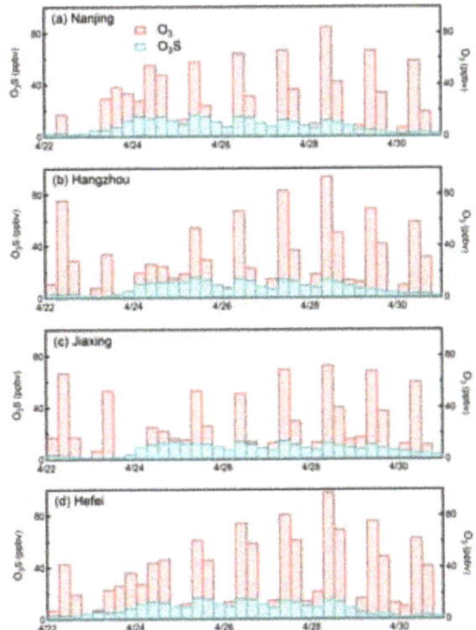

Figure 13. Time evolution of CAM-Chem simulated ozone (red) and O_3S (turquoise) in eastern China from 22 April 2018 to 30 April 2018. (**a**) Nanjing, (**b**) Hangzhou, (**c**) Jiaxing and (**d**) Hefei.

4. Conclusions and Discussion

Ozone pollution has been reported frequently and has become a serious issue in China [6,9,11]. Currently, most of the research has focused on anthropogenic and biogenic emitted ozone precursors from the surface [7,8,12], while the stratospheric ozone intrusion, another important source of tropospheric ozone, has been less studied. A recent study poses the question of to what extent do the SI events influence the surface ozone at relatively lower latitudes, that is, in eastern China [40]. This study is a follow-up work of a previous study [40], and investigates the potential contribution of SI events to surface ozone pollution episodes in eastern China during spring using ground-based measurements, the latest ERA5 reanalysis data as well as chemistry-climate model simulations.

Results indicate that a strong SI event occurred, associated with a horizontal-trough that brings ozone-rich air from the lower stratosphere to the surface and contributes to severe ozone pollution, during the period from 27 April to 28 April 2018 in eastern China. According to a CAM-Chem simulation, SI contributed about 15 ppbv (15%) to this surface ozone pollution episode in eastern China. Although such a contribution may not fully explain the mechanism of this severe ozone pollution episode, this study highlights the importance of SI events to the surface ozone variations. The potential impacts of SI events on tropospheric and surface ozone should be considered in ozone variation attribution and surface ozone forecasting.

Author Contributions: Y.W. initiated the study and did the analysis of meteorological conditions; H.W. analyzed the ozone transport. W.W. contributes to validation, project administration and wrote the original draft. All the authors contribute to interpretation, discussion, review and editing the manuscript. All authors have read and agreed to the published version of the manuscript.

Funding: This work was funded by the National Natural Science Foundation of China (Grant No. 41705023 and 41875095).

Acknowledgments: We thank the ECMWF for the ERA5 reanalysis data, China Ministry of Ecology and Environment for the ground-based ozone measurements and NCAR for the CAM-Chem simulations. We are grateful to the High Performance Computing Center (HPCC) of Nanjing University for doing the numerical calculations in this paper on its blade cluster system.

Conflicts of Interest: The authors declare no conflict of interest.

References

1. Stocker, T. *SBSTA-IPCC Special Event Climate Change 2013: The Physical Science Basis*; UNFCCC: New York, NY, USA, 2013.
2. Schultz, M.G.; Schröder, S.; Lyapina, O.; Cooper, O.; Galbally, I.; Petropavlovskikh, I.; von Schneidemesser, E.; Tanimoto, H.; Elshorbany, Y.; Naja, M.; et al. Tropospheric Ozone assessment report: Database and metrics data of global surface Ozone observations. *Elem. Sci. Anthr.* **2017**, *5*, 58–84. [CrossRef]
3. Mills, G.; Sharps, K.; Simpson, D.; Pleijel, H.; Frei, M.; Burkey, K.; Emberson, L.; Uddling, J.; Broberg, M.; Feng, Z.; et al. Closing the global ozone yield gap: Quantification and cobenefits for multistress tolerance. *Glob. Chang. Biol.* **2018**, *24*, 4869–4893. [CrossRef] [PubMed]
4. Young, P.J.; Naik, V.; Fiore, A.M.; Gaudel, A.; Guo, J.; Lin, M.Y.; Neu, J.L.; Parrish, D.D.; Rieder, H.E.; Schnell, J.L.; et al. Tropospheric Ozone Assessment Report:Assessment of global-scale model performance for global and regional ozone distributions, variability, and trends. *Elem. Sci. Anthr.* **2018**, *6*, 265.
5. Cooper, O.R.; Parrish, D.D.; Ziemke, J.; Cupeiro, M.; Galbally, I.E.; Gilge, S.; Horowitz, L.; Jensen, N.R.; Lamarque, J.F.; Naik, V.; et al. Global Distribution and Trends of Tropospheric Ozone: An Observation-Based Review. *Elem. Sci. Anthr.* **2014**, *2*, 29. [CrossRef]
6. Wang, T.; Xue, L.; Brimblecombe, P.; Lam, Y.F.; Li, L.; Zhang, L. Ozone pollution in China: A review of concentrations, meteorological influences, chemical precursors, and effects. *Sci. Total Environ.* **2017**, *575*, 1582–1596. [CrossRef]
7. Li, K.; Jacob, D.J.; Liao, H.; Shen, L.; Zhang, Q.; Bates, K.H. Anthropogenic drivers of 2013–2017 trends in summer surface ozone in China. *Proc. Natl. Acad. Sci. USA* **2019**, *116*, 422–427. [CrossRef]
8. Shen, L.; Jacob, D.J.; Liu, X.; Huang, G.; Li, K.; Liao, H.; Wang, T. An evaluation of the ability of the Ozone Monitoring Instrument (OMI) to observe boundary layer ozone pollution across China: Application to 2005–2017 ozone trends. *Atmos. Chem. Phys.* **2019**, *19*, 6551–6560. [CrossRef]
9. Ding, A.J.; Fu, C.B.; Yang, X.Q.; Sun, J.N.; Zheng, L.F.; Xie, Y.N.; Herrmann, E.; Nie, W.; Petäjä, T.; Kerminen, V.M.; et al. Ozone and fine particle in the western Yangtze River Delta: an overview of 1 yr data at the SORPES station. *Atmos. Chem. Phys.* **2013**, *13*, 5813–5830. [CrossRef]
10. Simpson, D.; Arneth, A.; Mills, G.; Solberg, S.; Uddling, J. Ozone—The persistent menace; interactions with the N cycle and climate change. *Curr. Opin. Environ. Sustain.* **2014**, *9*, 9–19. [CrossRef]
11. Xu, Z.; Huang, X.; Nie, W.; Shen, Y.; Zheng, L.; Xie, Y.; Wang, T.; Ding, K.; Liu, L.; Zhou, D.; et al. Impact of biomass burning and vertical mixing of residual-layer aged plumes on ozone in the Yangtze River Delta, China: A tethered-balloon measurement and modeling study of a multiday ozone episode. *J. Geophys. Res.* **2018**, *123*, 11786–11803. [CrossRef]
12. Wang, N.; Lyu, X.; Deng, X.; Huang, X.; Jiang, F.; Ding, A. Aggravating O_3 pollution due to NOx emission control in eastern China. *Sci. Total Environ.* **2019**, *677*, 732–744. [CrossRef] [PubMed]
13. Holton, J.R.; Haynes, P.H.; McIntyre, M.E.; Douglass, A.R.; Rood, R.B.; Pfister, L. Stratosphere-troposphere exchange. *Rev. Geophys.* **1995**, *33*, 403–439. [CrossRef]
14. Lelieveld, J.; Dentener, F.J. What controls tropospheric ozone. *J. Geophys. Res.* **2000**, *105*, 3531–3551. [CrossRef]
15. Ding, A.; Wang, T. Influence of stratosphere-to-troposphere exchange on the seasonal cycle of surface ozone at Mount Waliguan in western China. *Geophys. Res. Lett.* **2006**, *33*. [CrossRef]
16. Zhang, J.; Tian, W.; Xie, F.; Chipperfield, M.P.; Feng, W.; Son, S.W.; Abraham, N.L.; Archibald, A.T.; Bekki, S.; Butchart, N.; et al. Stratospheric ozone loss over the Eurasian continent induced by the polar vortex shift. *Nat. Commun.* **2018**, *9*, 206. [CrossRef]
17. Akritidis, D.; Katragkou, E.; Zanis, P.; Pytharoulis, I.; Melas, D.; Flemming, J.; Inness, A.; Clark, H.; Plu, M.; Eskes, H. A deep stratosphere-to-troposphere ozone transport event over Europe simulated in CAMS global and regional forecast systems: Analysis and evaluation. *Atmos. Chem. Phys.* **2018**, *18*, 15515–15534. [CrossRef]
18. Liang, Q.; Rodriguez, J.M.; Douglass, A.R.; Crawford, J.H.; Olson, J.R.; Apel, E.; Bian, H.; Blake, D.R.; Brune, W.; Chin, M.; et al. Reactive nitrogen, ozone and ozone production in the Arctic troposphere and the impact of stratosphere-troposphere exchange. *Atmos. Chem. Phys.* **2011**, *11*, 13181–13199. [CrossRef]

19. Emmons, L.K.; Hess, P.; Klonecki, A.; Tie, X.; Horowitz, L.; Lamarque, J.F.; Kinnison, D.; Brasseur, G.; Atlas, E.L.; Browell, E.; et al. Budget of tropospheric ozone during TOPSE from two chemical transport models. *J. Geophys. Res.* **2003**, *108*. [CrossRef]
20. Stohl, A.; Wernli, H.; James, P.; Bourqui, M.; Forster, C.; Liniger, M.A.; Seibert, P.; Sprenger, M. A new perspective of stratosphere-troposphere exchange. *Bull. Am. Meteorol. Soc.* **2003**, *84*, 1565–1573. [CrossRef]
21. Cooper, O.R.; Stohl, A.; Hübler, G.; Hsie, E.Y.; Parrish, D.D.; Tuck, A.F.; Kiladis, G.N.; Oltmans, S.J.; Johnson, B.J.; Shapiro, M.; et al. Direct transport of midlatitude stratospheric ozone into the lower troposphere and marine boundary layer of the tropical Pacific Ocean. *J. Geophys. Res.* **2005**, *110*. [CrossRef]
22. Akritidis, D.; Zanis, P.; Pytharoulis, I.; Mavrakis, A.; Karacostas, T. A deep stratospheric intrusion event down to the earth's surface of the megacity of Athens. *Meteorol. Atmos. Phys.* **2010**, *109*, 9–18. [CrossRef]
23. Langford, A.O.; Brioude, J.; Cooper, O.R.; Senff, C.J.; Alvarez, R.J.; Hardesty, R.M.; Johnson, B.J.; Oltmans, S.J. Stratospheric influence on surface ozone in the Los Angeles area during late spring and early summer of 2010. *J. Geophys. Res.* **2012**, *117*. [CrossRef]
24. Lin, M.; Fiore, A.M.; Cooper, O.R.; Horowitz, L.W.; Langford, A.O.; Levy, H.; Johnson, B.J.; Naik, V.; Oltmans, S.J.; Senff, C.J. Springtime high surface ozone events over the western United States: Quantifying the role of stratospheric intrusions. *J. Geophys. Res.* **2012**, *117*. [CrossRef]
25. Lefohn, A.S.; Wernli, H.; Shadwick, D.; Limbach, S.; Oltmans, S.J.; Shapiro, M. The importance of stratospheric–tropospheric transport in affecting surface ozone concentrations in the western and northern tier of the United States. *Atmos. Environ.* **2011**, *45*, 4845–4857. [CrossRef]
26. Lefohn, A.S.; Wernli, H.; Shadwick, D.; Oltmans, S.J.; Shapiro, M. Quantifying the importance of stratospheric-tropospheric transport on surface ozone concentrations at high- and low-elevation monitoring sites in the United States. *Atmos. Environ.* **2012**, *62*, 646–656. [CrossRef]
27. Yates, E.L.; Iraci, L.T.; Roby, M.C.; Pierce, R.B.; Johnson, M.S.; Reddy, P.J.; Tadić, J.M.; Loewenstein, M.; Gore, W. Airborne observations and modeling of springtime stratosphere-to-troposphere transport over California. *Atmos. Chem. Phys.* **2013**, *13*, 12481–12494. [CrossRef]
28. Xu, W.; Lin, W.; Xu, X.; Tang, J.; Huang, J.; Wu, H.; Zhang, X. Long-term trends of surface ozone and its influencing factors at the Mt Waliguan GAW station, China—Part 1: Overall trends and characteristics. *Atmos. Chem. Phys.* **2015**, *16*, 6191–6205. [CrossRef]
29. Li, D.; Bian, J.; Fan, Q. A deep stratospheric intrusion associated with an intense cut-off low event over East Asia. *Sci. China Earth Sci.* **2015**, *58*, 116–128. [CrossRef]
30. Zhang, J.; Tian, W.; Wang, Z.; Xie, F.; Wang, F. The Influence of ENSO on Northern Midlatitude Ozone during the Winter to Spring Transition. *J. Clim.* **2015**, *28*, 4774–4793. [CrossRef]
31. Baray, J.L.; Pointin, Y.; Baelen, J.V.; Lothon, M.; Campistron, B.; Cammas, J.P.; Masson, O.; Colomb, A.; Hervier, C.; Bezombes, Y.; et al. Case study and climatological analysis of upper tropospheric jet stream and stratosphere-troposphere exchanges using VHF profilers and radionuclide measurements in France. *J. Appl. Meteorol. Climatol.* **2017**, *56*, 3081–3097. [CrossRef]
32. Zhao, X.R.; Sheng, Z.; Li, J.W.; Yu, H.; Wei, K.J. Determination of the "wave turbopause" using a numerical differentiation method. *J. Geophys. Res.* **2019**, *124*, 10592–10607. [CrossRef]
33. Chen, X.L.; Ma, Y.M.; Kelder, H.; Su, Z.; Yang, K. On the behaviour of the tropopause folding events over the Tibetan Plateau. *Atmos. Chem. Phys.* **2010**, *11*, 22993–23016. [CrossRef]
34. Knowland, K.E.; Ott, L.E.; Duncan, B.N.; Wargan, K. Stratospheric Intrusion-Influenced Ozone Air Quality Exceedances Investigated in the NASA MERRA-2 Reanalysis. *Geophys. Res. Lett.* **2017**, *44*, 10691–10701. [CrossRef]
35. He, Y.; Sheng, Z.; He, M. Spectral Analysis of Gravity Waves from Near Space High-Resolution Balloon Data in Northwest China. *Atmosphere* **2020**, *11*, 133. [CrossRef]
36. Chen, X.; Añel, J.A.; Su, Z.; de la Torre, L.; Kelder, H.; van Peet, J.; Ma, Y. The Deep Atmospheric Boundary Layer and Its Significance to the Stratosphere and Troposphere Exchange over the Tibetan Plateau. *PLoS ONE* **2013**, *8*. [CrossRef]
37. Chen, D.; Lü, D.; Chen, Z. Simulation of the stratosphere-troposphere exchange process in a typical cold vortex over Northeast China. *Sci. China Earth Sci.* **2014**, *57*, 1452–1463. [CrossRef]
38. Liang, M.C.; Mahata, S. Oxygen anomaly in near surface carbon dioxide reveals deep stratospheric intrusion. *Sci. Rep.* **2015**, *5*, 11352. [CrossRef]

39. Zhang, J.; Tian, W.; Xie, F.; Tian, H.; Luo, J.; Zhang, J.; Liu, W.; Dhomse, S. Climate warming and decreasing total column ozone over the Tibetan Plateau during winter and spring. *Tellus B Chem. Phys. Meteorol.* **2014**, *66*, 23415. [CrossRef]
40. Wang, H.; Wang, W.; Huang, X.; Ding, A. Impacts of stratosphere-to-troposphere-transport on summertime surface ozone over eastern China. *Chin. Sci. Bull.* **2019**. [CrossRef]
41. Xu, Z.; Huang, X.; Nie, W.; Chi, X.; Xu, Z.; Zheng, L.; Sun, P.; Ding, A. Influence of synoptic condition and holiday effects on VOCs and ozone production in the Yangtze River Delta region, China. *Atmos. Environ.* **2017**, *168*, 112–124. [CrossRef]
42. Copernicus Climate Change Service (C3S), ERA5: Fifth Generation of ECMWF Atmospheric Reanalyses of the Global Climate. Copernicus Climate Change Service Climate Data Store (CDS). 2017. Available online: https://cds.climate.copernicus.eu/cdsapp#!/home (accessed on 29 January 2020).
43. Emmons, L.K.; Hess, P.G.; Lamarque, J.F.; Pfister, G.G. Tagged ozone mechanism for MOZART-4, CAM-chem and other chemical transport models. *Geosci. Model Dev.* **2012**, *5*, 1531–1542. [CrossRef]
44. Stein, A.F.; Draxler, R.R.; Rolph, G.D.; Stunder, B.J.B.; Cohen, M.D.; Ngan, F. NOAA's HYSPLIT Atmospheric Transport and Dispersion Modeling System. *Bull. Am. Meteorol. Soc.* **2015**, *96*, 2059–2077. [CrossRef]
45. Ding, A.; Wang, T.; Fu, C. Transport characteristics and origins of carbon monoxide and ozone in Hong Kong, South China. *J. Geophys. Res.* **2013**, *118*, 9475–9488. [CrossRef]

Article

Assessing the Impact of Ozone and Particulate Matter on Mortality Rate from Respiratory Disease in Seoul, Korea

Sun Kyoung Park

School of ICT-Integrated Studies, Pyeongtaek University, Pyeongtaek 17869, Korea; skpark@ptu.ac.kr

Received: 13 October 2019; Accepted: 5 November 2019; Published: 7 November 2019

Abstract: The evidence linking ozone and particulate matter with adverse health impacts is increasing. The goal of this study was to assess the impact of air pollution on the mortality rate from respiratory disease in Seoul, Korea, between 2008 and 2017. The analysis was conducted using a decision tree model in two ways: using 24-h average concentrations and using 1-h maximum values to compare any health impacts from the different times of exposure to pollution. Results show that in spring an elevated level of ozone is one of the most important factors, but in summer temperature has a greater impact than air pollution. Nitrogen dioxide is one of the most important factors in fall, while high levels of particles less than 2.5 μm ($PM_{2.5}$) and 10 μm in size (PM_{10}) and cooler temperatures are key factors in winter. We checked the accuracy of our results through a 10-fold cross validation method. Error rates using 24-h average and 1-h maximum concentrations were in the ranges of 24.9–42% and 27.6–42%, respectively, indicating that 24-h average concentrations are slightly more directly related with mortality rate. These results could be useful for policy makers in determining the temporal scale of predicted pollutant concentrations for an air quality warning system to help minimize the adverse impacts of air pollution.

Keywords: ozone; $PM_{2.5}$; PM_{10}; nitrogen dioxide; respiratory disease; decision tree model

---

1. Introduction

Air pollution reduces visibility and has an adverse impact on human health [1]. Relatively high pollutant levels are often observed in industrialized cities. As a representative example of air pollutants, the tropospheric ozone (O_3) is a secondary pollutant produced naturally by photochemical decomposition. O_3 maintains an equilibrium concentration between production and removal, but artificially emitted nitrogen dioxide (NO_2) and volatile organic carbons (VOCs) accelerate the production of O_3 through a photochemical reaction. Accordingly, high O_3 concentrations are common in cities, which damage ecosystems [2–4]. Prolonged exposure to high O_3 concentrations causes or exacerbates cardiovascular disease and respiratory diseases, such as pneumonia, chronic obstructive pulmonary disease, asthma, and allergic rhinitis. Cases leading to death have also been reported [5–7].

Particulate matter is also a representative air pollutant with direct effects on human health. Particles less than 2.5 μm in size are referred to as $PM_{2.5}$, and particles less than 10 μm in size referred to as PM_{10}. Because inhaled fine particles can penetrate deep into the capillary vessels, particulate matter is known as a direct cause of cardiovascular disease. There have been multiple reports showing that exposure to high particle concentrations leads to increased fetal mortality. According to the World Health Organization (2014), by 2012, the global toll of premature deaths related to air pollution had reached 7 million people annually. As such, numerous research results showing air pollution's direct and indirect impacts on human health have been published [8].

Air pollution affects health in many ways [9–11]. The impact of air pollution on school students was studied for 3.5 years in Barcelona, Spain [12]. Results showed that an increase in ambient NO_2

and particulate matter concentrations by one interquartile range deteriorated memory development in students by around 20%. Ljungman et al. (2018) analyzed the relationship between air pollution and arterial stiffness. Although the result showed no linkage between arterial stiffness and $PM_{2.5}$, a higher probability of arterial stiffness was found in roadside residents. In other words, although the impact of a single type of air pollution was not clear, the results demonstrated that several air pollutants have complex effects on health [13]. Therefore, the influence of multiple air pollutants on the human body should be analyzed to accurately identify the effect of air pollution.

The impact of air pollution on human health also varies according to the analytical method used. Son et al. (2010) analyzed the impact of air pollution on pulmonary function by using PM_{10}, O_3, NO_2, SO_2, and CO obtained from 13 observatories in Ulsan, Korea, from 2003 to 2007 [11]. Four methods were used to calculate representative pollutant concentrations: simple averaging, nearest distance, inverse distance weighting, and kriging. Subsequent tests of the accuracy of the analysis determined that kriging was superior. As such, differences in research results may occur depending on the method selected. On the other hand, results may vary depending on the timescale of the observed air pollution. Lee et al. (2018) analyzed the impact of short-term exposure to air pollution (fewer than 8 days) and long-term exposure to air pollution (annual) on key inflammatory markers by using linear mixed effects models [14]. The results showed that although short-term exposure was related to increased fibrinogen and ferritin levels, long-term exposure was related to fibrinogen and white blood cell counts. Likewise, the impact of air pollution may vary depending on the temporal scale of data, so it is meaningful to compare epidemiological study results using data of different timescales.

The purpose of this study is two-fold: one reason is to find pollutant levels that determine the high probability of mortality from respiratory disease, and the other is to compare the accuracy of the results using 1-h maximum with that using 24-h average pollutant concentrations. In order to achieve these goals, we classified the dependent variable as days with high and low probability of mortality. In addition, the effect of temperature on health was also analyzed because temperature is known to have a direct influence on health [6,10]. The statistical model used was a decision tree algorithm. Among various statistical models, the decision tree algorithm was especially useful for finding factors with which to classify the dependent variable [15]. A brief description of the model is provided in the next section.

2. Research Methods

2.1. Analytical Data

Hourly air pollutant concentrations, temperature, and daily number of deaths in Seoul from 2008 to 2017 were used for this study. Air pollution data for SO_2, CO, O_3, NO_2, PM_{10}, and $PM_{2.5}$ were measured from 25 monitoring stations operated by the Korea Environment Corporation (KECO) (Figure 1) [16]. Temperatures were collected from the Korea Meteorological Administration's (KMA) National Climate Data Center (NCDC) [17]. The number of deaths was based on the public microdata of the National Statistical Office (NSO) [18]. The number of deaths from respiratory disease was in the "J00–J99" category of the 10th International Classification of Diseases (ICD-10).

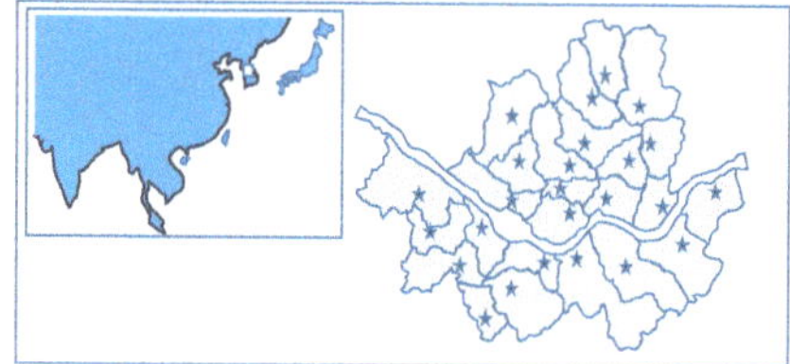

Figure 1. Air pollution monitoring stations in Seoul, Korea.

2.2. Decision Tree Model

Decision tree models can efficiently accommodate any data formats that are non-normal, a mix of continuous, discrete, and categorical formats, and cross- or auto-correlated formats. Moreover, the decision trees facilitate the interpretation of the final model because their output is a hierarchical structure that consists of a series of "if–then" rules to predict the outcome of the dependent variable [19]. This cannot be easily achieved using other time series regression models, such as distributed lag non-linear function. A decision tree model expresses rules appropriate for classifying or predicting dependent variables (i.e., number of deaths caused from respiratory disease) based on independent variables (i.e., air pollutant concentration, etc.). Here, independent variables were the 1-h maximum and 24-h average air pollutant concentrations and 24-h average temperatures. Dependent variables were the categorical values of days with high numbers of deaths (H) or days with low numbers of deaths (L) and were classified based on the median number of deaths from respiratory disease (Table 1).

A classification and regression tree (CART) was used to apply the decision tree model. Because detailed descriptions of CART models can be found in other literature, only a brief description is outlined in this paper [15]. First, CART makes classifications based on the most important independent variable (Figure 2). For example, let us assume independent variables are X, Y, and Z. Here, the first basis for classifying dependent variables as A or B is "$X \leq x$". The second basis is "$Y \leq y$" or "$Z \leq z$". Under "$X \leq x$", the dependent variable is classified as A if "$Y \leq y$", and it is classified as B if "$Y > y$". Under "$X > x$", the dependent variable is classified as B if "$Z \leq z$", and it is classified as A if "$Z > z$". Among the independent variables, only appropriate variables are used for the classification.

The classification criteria in CART maximize similarity and dissimilarity among groups based on the Gini index [15]. Because the branch divides repeatedly based on this method, a tree-shaped structure is hierarchically constructed. the visual expression is helpful in understanding and interpreting the results. The model has been widely used in many areas, including environmental sciences and epidemiological studies linking air pollution and human health [20–26].

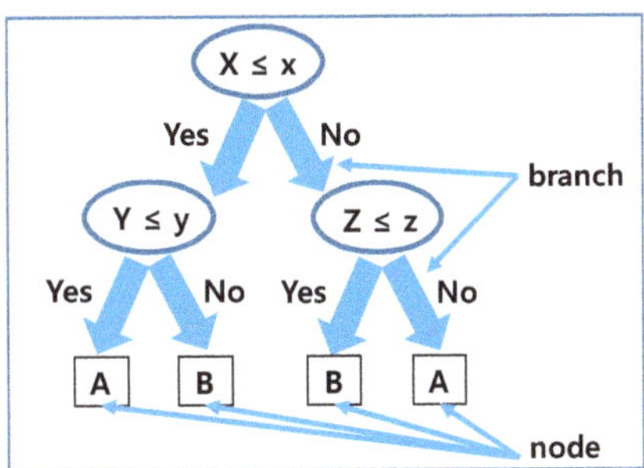

Figure 2. Schematic diagram of classification. Independent variables are X, Y, and Z, and dependent variables are classified as A and B.

The dependent variable in the model was categorized as days with relatively high (H) and low (L) numbers of daily deaths compared with the median number of daily deaths from respiratory disease. The median number of daily deaths during the analysis period was seven deaths. Thus, "H" indicates days where the number of deaths from respiratory disease is higher than seven, and "L" indicates days where the number of deaths is equal to or lower than seven.

Independent variables included the 1-h maximum pollutant concentrations (SO_2, CO, O_3, PM_{10}, $PM_{2.5}$, and NO_2) and the 24-h average temperature. Air pollutant concentrations one to three days before the deaths were included in the independent variables to observe the influence from pollution prior to death. Temperature is known to have a relatively long-term effect compared with air pollution [10]. Therefore, temperatures from four to 20 days before the deaths were also included in the independent variables. To decrease the number of independent variables, average temperatures from four to 10 days and from 11 to 20 days were used instead of using temperatures on each day. A CART model's accuracy does not increase even though the number of independent variables is increased if there is correlation between the independent variables [15]. The 1-h air pollutant concentration n days before the deaths is abbreviated as "Pollutant name"_max(n d) hereafter. For example, "O_3_max(1 d)" indicates the 1-h maximum O_3 concentration, one day before death. The average temperatures from 4 to 10 days and from 11 to 20 days are represented as "T(4–10 d)" and "T(11–20 d)", respectively.

Taking into account the annual changes in air pollutant concentrations, an analysis was conducted for each month. For example, O_3 concentration was especially high in May and June, and prolonged exposures to high ozone concentrations could be a direct cause of respiratory and eye diseases. On the other hand, particle concentrations were relatively high in winter. As such, it was difficult to accurately identify the factors causing mortality without adding seasonal distinctions into the analysis. Previous studies that used decision tree models to conduct studies related to air pollution also limited the analyses periods to several weeks and separate analyses were conducted for each season. For example, Chu et al. (2012) limited the analysis period to spring (from 28 April to 13 May 2009) to find factors that influenced ozone concentrations [24]. Park (2018) constructed independent models for each season to assess factors linked with cardiovascular disease [27].

3. Status of Air Pollution, Temperature, and the Number of Deaths Caused by Air Pollutants

3.1. Air Pollution and Temperature

Basic statistics of hourly SO_2, CO, O_3, NO_2, PM_{10}, and $PM_{2.5}$ concentrations in Seoul, Korea, from 2008 and 2017 are illustrated as box plots in Appendix A (Figure A1). The plots show that SO_2 and CO met the air quality standard at all times. Unsurprisingly, O_3 exceeded the air quality standard from April to September; because strong sunlight accelerates O_3 generation; as such, O_3 generally increases in spring and summer [28–30]. The 24-h average NO_2 exceeded the air quality standard from January to May and from October to December. Moreover, the 1-h average NO_2 exceeded the standard regardless of the season. Much like NO_2, the 24-h PM_{10} also exceeded the standard during the relatively cold seasons. One of the important causes of the higher PM_{10} concentrations during the cold seasons was the relatively low atmospheric mixing height because of a low ground temperature [31]. The year-round exceedance of $PM_{2.5}$ shows how imperative it is to reduce $PM_{2.5}$. Daily average temperatures showed clear seasonal changes; the highest temperature was 33.7 °C in August and the lowest value was −14.8 °C in January.

3.2. Number of Deaths Caused by Respiratory Disease

The number of deaths in Korea from 2008 to 2017 was around 2.6 million, so the annual average number of deaths was approximately 260,000. Differences in the number of deaths varied up to 15%, depending on the months. The mortality rate was relatively high in summer and winter, and relatively low in the spring and fall. Causes of death were cancers (28%), cardiovascular disease (22%), traffic accidents and suicides (11%), diabetes and liver disease (9%), respiratory disease (28%), and others (22%). DeLeon and Thruston (2003) found that the influence of air pollution on deaths was clear for the elderly, but less clear for others [32]. Accordingly, this study also focuses on persons aged 65 or older at time of death from respiratory disease (Table 1).

Table 1. Basic statistics of the daily number of deaths from respiratory disease for persons aged 65 or older at time of death from 2008 to 2017 in Seoul, Korea.

Mean	Standard Deviation	Median	Maximum	Minimum
7.6	3.5	7	23	1

4. Results and Discussion

4.1. Linkage of Air Pollution and Temperature with Mortality from Respiratory Disease

Monthly average pollutant concentrations on days with a higher probability of deaths from respiratory disease were compared with those with lower probability of deaths to find the linkage between air pollution and deaths (Figures 3–7). Differences in O_3 concentrations were relatively large from May to August when high O_3 concentrations were observed (Figure 3). Studies have shown that the adverse health effects from O_3 are often found in industrialized cities, in which the production of O_3 is accelerated by NOx and VOCs emissions [4–7]. Because strong sunlight is crucial for O_3 production, high O_3 levels are often found in spring, as was true in this case (Figure 3).

The health impact of air pollution may appear within a few hours after exposure to pollution or several days afterward [5]. Consequently, a time delay may exist between the occurrence of high air pollution and death. To take this possibility into account, O_3 concentrations up to five days before the day of death were analyzed; data in February, May, August, and November are presented as an example (Figure 3).

The O_3 concentrations up to five days before recorded deaths were obviously higher than on days with lower numbers in May (Figure 3). The 1-h maximum O_3 values in May on the days with high and low numbers of deaths were 86 ppb and 78.3 ppb, respectively, with a difference of 7.7 ppb (Figure 3a). Differences at 1, 2, 3, 4, and 5 days before death were 8.2 ppb, 8.9 ppb, 4.3 ppb, 6.8 ppb, and 5.6 ppb, respectively, indicating that the difference at 1–2 days before death was greater than that on the day of death. The 8-h average O_3 values in May on days with high and low numbers of deaths were 72.9 ppb and 66.4 ppb, respectively, with a difference of 6.4 ppb (Figure 3b). Differences at 1, 2, 3, 4, and 5 days before death were 6.5 ppb, 6.6 ppb, 3.1 ppb, 5.0 ppb, and 4.1 ppb, respectively. The results indicated that O_3 concentrations 1–2 days before death had a direct association with death, whereas O_3 concentration 3–5 days before death had relatively less impact.

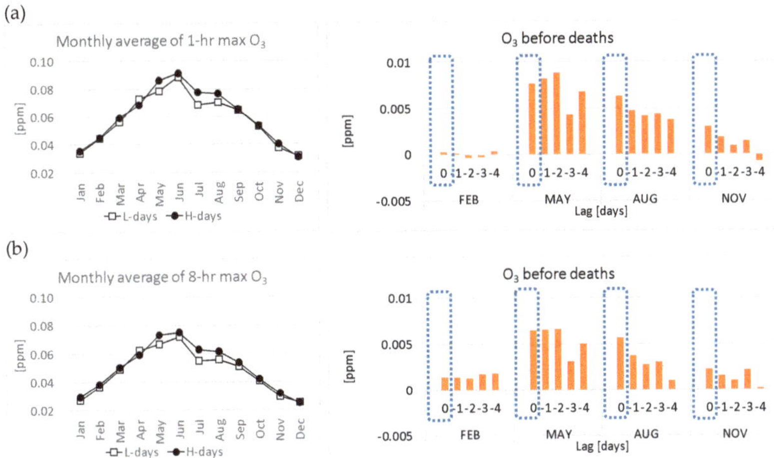

Figure 3. Ozone (O_3) concentrations on days with a higher or lower probability of deaths from respiratory disease, and those from zero to 5 days before death in February, May, August, and November between 2008 and 2017: (**a**) 1-h and (**b**) 8-h maximum O_3 concentrations.

$PM_{2.5}$ concentrations did not show clear differences between the days with high and low numbers of deaths (Figure 4). Differences in $PM_{2.5}$ concentrations in August did not consistently increase or decrease on days before death. However, differences 1 day and 2 days before death increased in February, May, and November, while differences decreased 3, 4, and 5 days before death. These results imply that high $PM_{2.5}$ 1 day or 2 days before death is associated with deaths from respiratory disease, which is consistent with an existing study showing that deaths from cardiovascular disease occur a few days after high $PM_{2.5}$ concentrations [33].

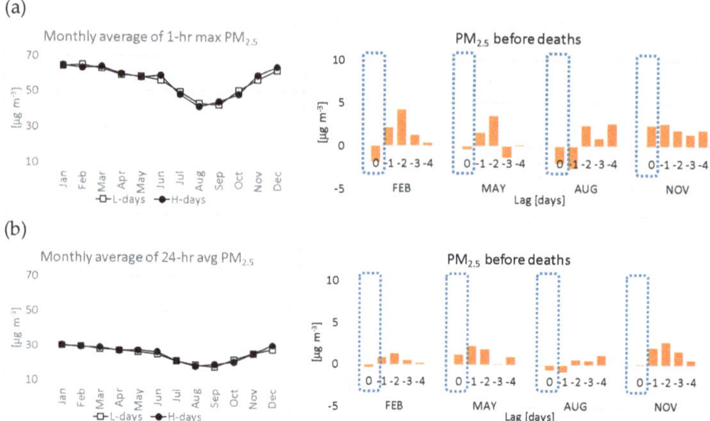

Figure 4. Concentrations of particles less than 2.5 μm in size ($PM_{2.5}$) on days with a higher or lower probability of deaths from respiratory disease, and those from zero to 5 days before deaths in February, May, August, and November between 2008 and 2017: (**a**) 1-h maximum and (**b**) 24-h average $PM_{2.5}$ concentrations.

PM_{10} concentrations were relatively higher in winter and in spring (Figure 5). The high PM_{10} concentrations during relatively cold seasons are also related to the relatively low mixing height during the cold season, because pollutants can accumulate in the lower troposphere if the mixing height is low [34]. The period in which maximum PM_{10} concentrations were observed was consistent with days with an inflow of yellow dust from the west of Korea [35]. Differences in PM_{10} concentrations 1 day and 2 days before death were obvious in February, May, and November, which implied that PM_{10} concentrations one or two days before death had a large impact.

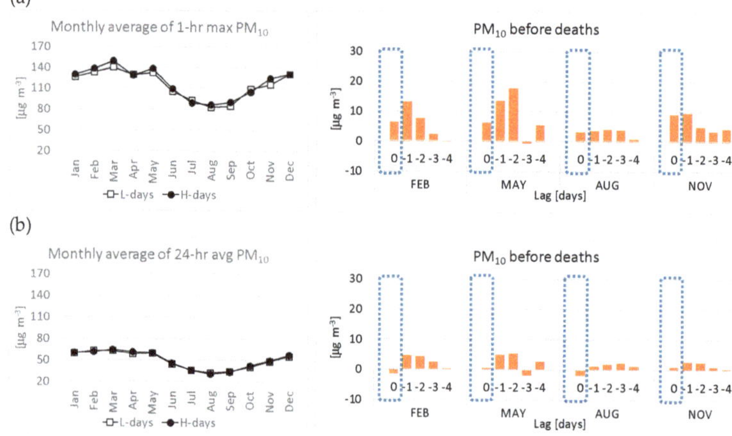

Figure 5. PM_{10} concentrations on days with a higher or lower probability of deaths from respiratory disease, and those from zero to 5 days before deaths in February, May, August, and November between 2008 and 2017: (**a**) 1-h maximum and (**b**) 24-h average PM_{10} concentrations.

Monthly average NO_2 concentrations on days with high numbers of deaths did not show a big difference to those with low numbers of deaths (Figure 6). NO_2 concentrations from one to five days before days with high numbers of deaths differed little from those with low numbers of deaths. Such results do not necessarily mean that NO_2 concentration was not a direct cause of death from respiratory disease, because the recorded monthly average concentration alone is insufficient for a thorough analysis of its effects. Accordingly, a decision tree model was introduced to permit closer observations of the health impacts of air pollution.

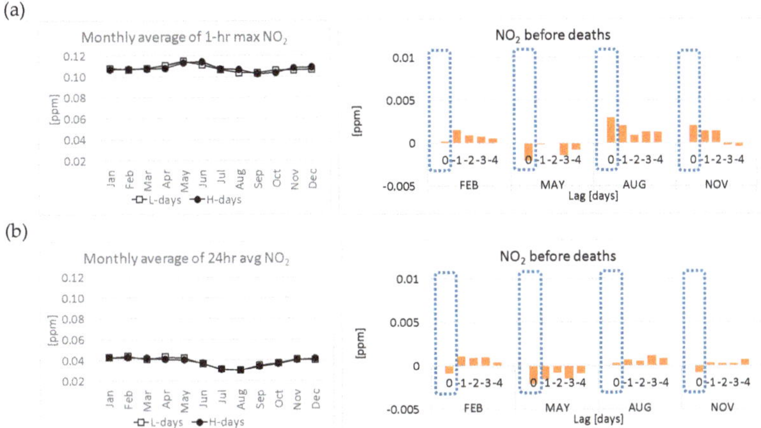

Figure 6. Nitrogen dioxide (NO_2) concentrations on days with a higher or lower probability of death from respiratory disease and those from zero to 5 days before deaths in February, May, August, and November between 2008 and 2017: (**a**) 1-h maximum and (**b**) 24-h average NO_2 concentrations.

The average temperature on days from February to April with high numbers of deaths was lower than on days with a low number of deaths (Figure 7). This result is consistent with a previous study showing that high numbers of deaths by respiratory disease occurred on days with low temperatures in winter [36,37]. However, average temperatures on the days with high numbers of deaths were slightly higher in most cases than on the days with low numbers of deaths. This was partly because of the relatively low particle concentrations observed on cold days.

Air pollutant concentrations, especially those of secondary pollutants such as O_3, are affected by meteorological conditions [29]. Moreover, when cold air masses move in from the relatively clean air of the northern polar area, particle concentrations tend to be low. Similar phenomena are observed in Vietnam as well. Hien et al. (2011) observed a reduced PM_{10} concentration from October to February in Hanoi, Vietnam, immediately before a cold surge occurred [38]. Similarly, temperature and air pollution are related. As a result, it is necessary to simultaneously analyze the effect of both air pollution and temperature on health to isolate the factors exacerbating respiratory disease.

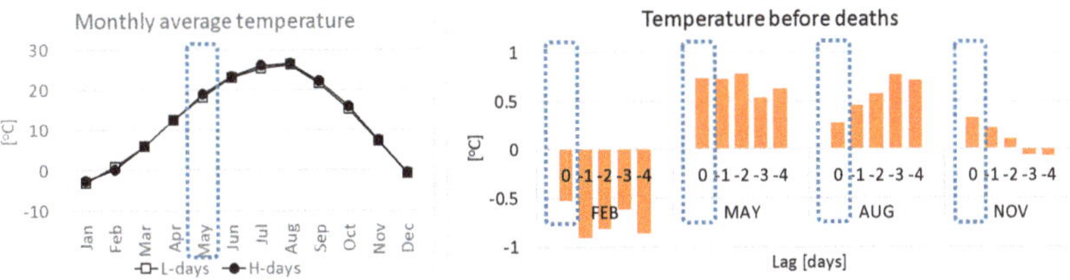

Figure 7. Temperatures on days with a higher or lower probability of deaths from respiratory disease, and temperatures from zero to 5 days before deaths in May between 2008 and 2017.

4.2. Factors that Impact the Number of Deaths Caused by Respiratory Disease

4.2.1. Influence of 1-h Maximum Pollutant Concentrations on the Number of Deaths from Respiratory Disease

The decision tree model was used to observe the effect of pollutant concentrations on the number of deaths caused by respiratory disease. Originally, the model was constructed using all data, and the accuracy of the model was checked by the 10-fold cross validation error. Results showed that the error rate was 45%, indicating that the model did not accurately classify factors affecting high and low probability of death from respiratory disease, partly due to seasonal variations of pollutant concentrations. Thus, the analyses were conducted separately for each month.

The impact of air pollution and temperature on respiratory disease as analyzed with a CART model could be interpreted as follows (Figure 8). Here, "T(11–20 d)" became the basis for the first branch in January. This signified that the most important factor that impacted the risk of death was "T(11–20 d)". Results showed that the risk of death was relatively low when "T(11–20 d)" was higher than 2.4 °C. When "T(11–20 d)" was 2.4 °C or lower, "$PM_{2.5}$_max(1 d)", the basis of the next branch, was checked to determine the risk of death. When "T(11–20 d)" was less than 2.4 °C and "$PM_{2.5}$_max(1 d)" was higher than 95.5 $\mu g \cdot m^{-3}$, the risk of death was relatively high. If the "$PM_{2.5}$_max(1 d)" was 95.5 $\mu g \cdot m^{-3}$ or less, "PM_{10}_max(1 d)" was checked. If "PM_{10}_max(1 d)" exceeded 125.5 $\mu g \cdot m^{-3}$, the risk of death was relatively high. Assuming that "PM_{10}_max(1 d)" was 125.5 $\mu g \cdot m^{-3}$ or less, if the risk of death was relatively high when "NO_2_max(1 d)" was higher than 83 ppb, the risk of death was relatively low if "NO_2_max(1 d)" was 83 ppb or less. Based on this, it was possible to analyze the linkage between temperature, PM_{10}, $PM_{2.5}$, and NO_2 concentrations with the risk of death.

Here, "T(4–10 d)" was the most important factor in February. When "T(4–10 d)" exceeded 3.8 °C, the risk of death was relatively low. On the other hand, when "T(4–10 d)" was 3.8 °C or less, "T(3 d)" was checked to ascertain the risk of death. Even if "T(4–10 d)" was less than 3.8 °C, the risk of death was relatively low when "T(3 d)" was higher than 5.7 °C. However, when "T(4–10 d)" was less than 3.8 °C and "T(3 d)" was less than 5.7 °C, the risk of death differed depending on "PM_{10}_max(1 d)". Although the risk of death also increased when "PM_{10}_max(1 d)" was higher than 143 $\mu g \cdot m^{-3}$, it was relatively low when "PM_{10}_max(1 d)" was 143 $\mu g \cdot m^{-3}$ or below. Such results showed that "T(4–10 d)" was the most direct factor and that it was also important in February.

High PM_{10} concentrations were frequently observed, partly because of yellow dust in March (Figure A1e) [39,40]. "T(2 d)", "PM_{10}_max(1 d)", and "NO_2_max(1 d)" were among the important factors in March. "NO_2_max(1 d)", "$PM_{2.5}$_max(1 d)", and "PM_{10}_max(1 d)" were related to the risk of death. Those results were consistent with previous studies that showed NO_2 and particulate matter were directly related to deaths from respiratory disease. Dong et al. (2012) confirmed through a study in Shenyang, China, that the risk of death from respiratory disease increased when PM_{10} and NO_2 concentrations were relatively high [41].

The 1-h maximum O_3 concentrations before the deaths under study were the most important factors in May and in June. Burnet et al. (1997) illustrated the association between O_3 concentrations one day before hospitalization and the number of hospitalized patients by analyzing patients in 16 cities in Canada from April 1981 to December 1991 [42]. Although "T(2 d)" was the most important factor in July, "O_3_max(1 d)" was also closely related with the deaths. Reid et al. (2012) found that the risk of death increased when both O_3 concentrations and temperatures were high [6].

The temperatures 1, 2, and 3 days before the deaths were linked with the cause of death in August (Figure 8). This result was consistent with a previous study that showed that the risk of death increased with higher temperatures in August [43]. "NO_2_max(1 d)" was the most important factor relating to risk of death in September and in October. High PM_{10} and $PM_{2.5}$ concentrations were frequently observed in November and in December, and "T(11–20 d)", "PM_{10}_max(1 d)", and "$PM_{2.5}$_max(1 d)" were important factors that determined the risk of death from respiratory disease.

The accuracy of results was ascertained through 10-fold cross validation of errors [44,45] (Table 2). Error rates were 27.6–42%, with the highest value in August. Errors were also relatively higher than for a similar study that analyzed factors influencing the number of deaths caused by cardiovascular disease [27]. The risk of death from respiratory disease was determined by only the temperature before deaths in August because pollutant levels were relatively low (Figure A1). One of the causes of the low pollutant concentrations in summer was the dilution of air pollution by the elevated mixing height [31]. In addition, the relatively high precipitation in summer inhibited the photochemical formation of O_3 and accelerated the wet deposition of particles [46,47]. Accordingly, the risk of death was determined by only temperature in August, so the accuracy of the decision tree model may have been reduced.

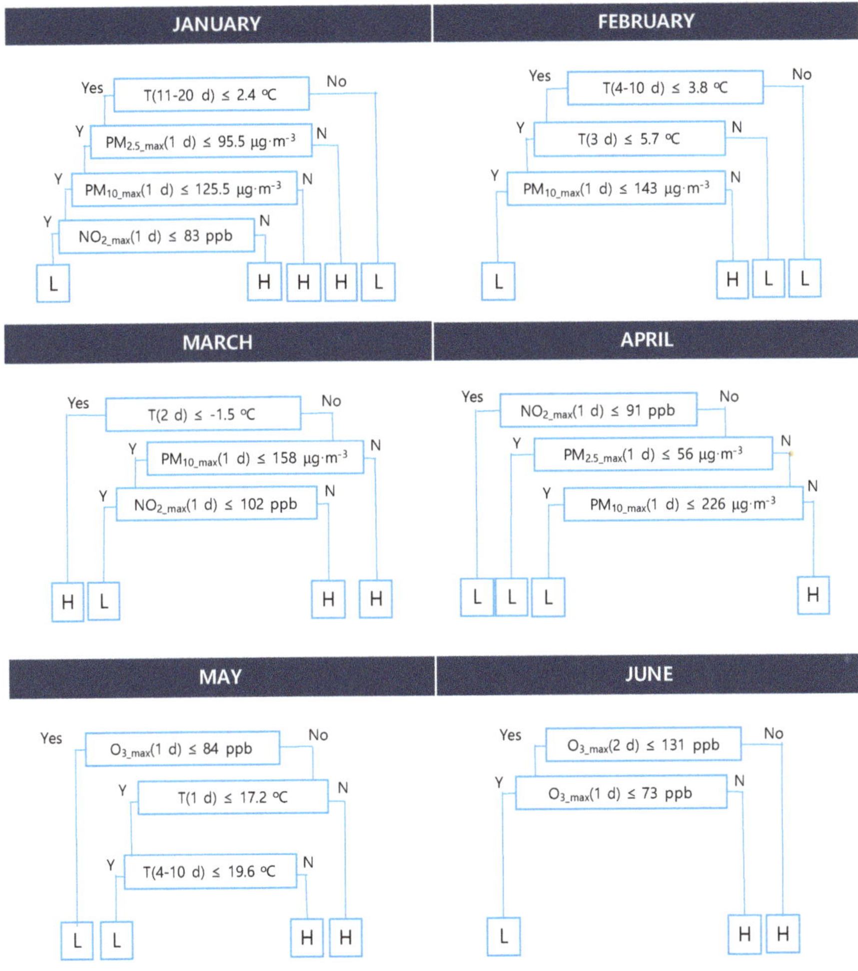

Figure 8. *Cont.*

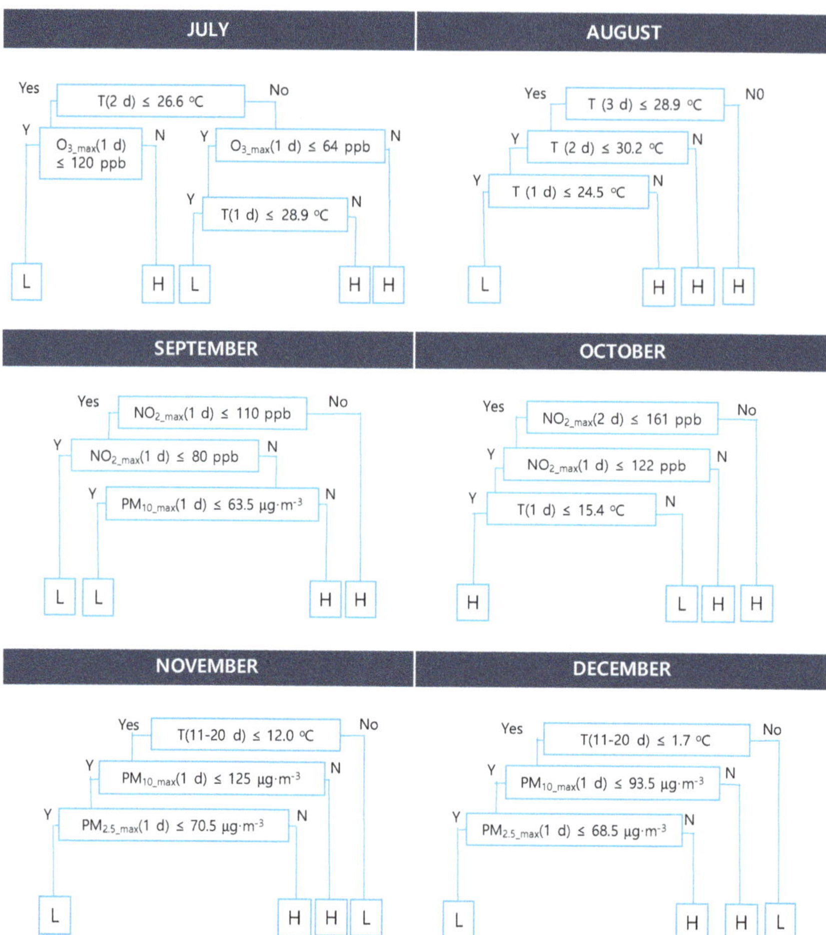

Figure 8. Hourly maximum pollutant concentrations and temperature affecting the high (H) and low (L) probability of death from respiratory disease.

4.2.2. Influence of 24-h Average Pollutant Concentrations on the Number of Deaths from Respiratory Disease

The health impact of air pollution varied depending on how long people were exposed [13,14]. Zanobett et al. (2003) analyzed the influence of PM_{10} up to 40 days after exposure [48]. Results showed that when PM_{10} concentrations increased by 10 μg·m^{-3}, the risk of death from respiratory disease increased by 0.74%. The risk of death increased by five times if the exposure to air pollution lasted for more than a month. Therefore, to compare the effect of different exposure times, we analyzed the influence of 24-h average pollutant concentrations on deaths. Thus, the impact of the 24-h average pollutant levels was compared with that of the 1-h maximum pollutant concentrations analyzed in the previous section (Figure 9). The 24-h pollutant concentration n days before death is abbreviated as "Pollutant name"_avg(N d) hereafter. For example, "O_3_avg(1 d)" indicates the 24-h average O_3 concentration one day before death.

The first branch that classified the high and low risk of death in January was "T(11–20 d)". The second branch was classified based on "PM_{10}_avg(2 d)". The risk of death was relatively low when "T(11–20 d)" was 2.4 °C or lower. When "T(11–20 d)" was higher than 2.4 °C, the risk of death was high if "PM_{10}_avg(2 d)" was higher than 36.5 $\mu g \cdot m^{-3}$. When "PM_{10}_avg(2 d)" was less than 36.5 $\mu g \cdot m^{-3}$, the risk of death was relatively high if "$PM_{2.5}$_avg(2 d)" exceeded 29.4 $\mu g \cdot m^{-3}$. When "$PM_{2.5}$_avg(2 d)" was less than 29.4 $\mu g \cdot m^{-3}$, the risk of death was relatively high if "NO_2_avg(2d)" was higher than 52 ppb.

"T(4–10 d)" was directly related in February with the risk of death because of respiratory disease. When "T(4–10 d)" exceeded 3.8 °C, the risk of death was relatively low. When "T(4–10 d)" was less than 3.8 °C, the risk of death was relatively low only if "PM_{10}_avg(2 d)" exceeded 52.8 $\mu g \cdot m^{-3}$. When "PM_{10}_avg(2 d)" was less than 52.8 $\mu g \cdot m^{-3}$, the risk of death was classified as relatively high if "$PM_{2.5}$_avg(2 d)" exceeded 25 $\mu g \cdot m^{-3}$, but the risk was relatively low if "$PM_{2.5}$_avg(2 d)" was lower than 25 $\mu g \cdot m^{-3}$. Likewise, "T(4–10 d)", "PM_{10}_avg(2 d)", and "$PM_{2.5}$_avg(2 d)" were major factors in February.

"T(2 d)", "PM_{10}_avg(2 d)", and "$PM_{2.5}$_avg(2 d)" were factors in March that influenced the number of deaths from respiratory disease. Factors associated with the risk of death in April included "NO_2_avg(2 d)", "PM_{10}_avg(1 d)", and "$PM_{2.5}$_avg(1 d)". High O_3 concentrations were often observed in May. "O_3_avg(1 d)" and temperatures before death were among the important factors. "O_3_avg(2 d)" and "T(1 d)" were important in determining the risk of death.

Although "T(2d)" was the most important factor in July, "T(1 d)" and "O_3_avg(1 d)" were also associated with deaths as well. The risk of death increased along with the higher temperatures in August. "NO_2_avg(2 d)" was the most important factor related to the risk of death in September and in October. However, "T(11–20 d)" and "PM10_avg(1 d)" were the important risk factors in November and in December.

The 10-fold cross validation errors resulted in 24.1–42% of errors, depending on the month (Table 2). The errors were greater than those in a similar study that analyzed the relation between air pollution and cardiovascular disease [27]. We also conducted the analysis using a 3-day cumulative pollutant. Factors linked with the death from respiratory disease using a 3-day cumulative pollutant were similar to those using separate data on each day. However, the accuracy of the model using a 3-day cumulative pollutant was slightly worse than that using daily pollutant concentrations.

Errors analyzed using the 24-h average concentrations were slightly less than those analyzed using the 1-h maximum concentrations. The result indicated that the 24-h average concentrations were more directly related to the risk of death than 1-h maximum concentrations. Jerrett et al. (2004) confirmed differences in degrees of exposure leading to differences in mortality through research conducted in Hamilton, Canada [49]. If the subject of the analysis did not engage in outdoor activities during the period in which the 1-h maximum concentration was observed, a direct association between high pollutant concentrations and the risk of death did not occur, even though the 1-h maximum concentration was directly associated with health. Therefore, the results of this study alone should not be interpreted as proving that short-term exposure to extreme levels of pollution was relatively less hazardous to health than long-term exposure to elevated levels of air pollution.

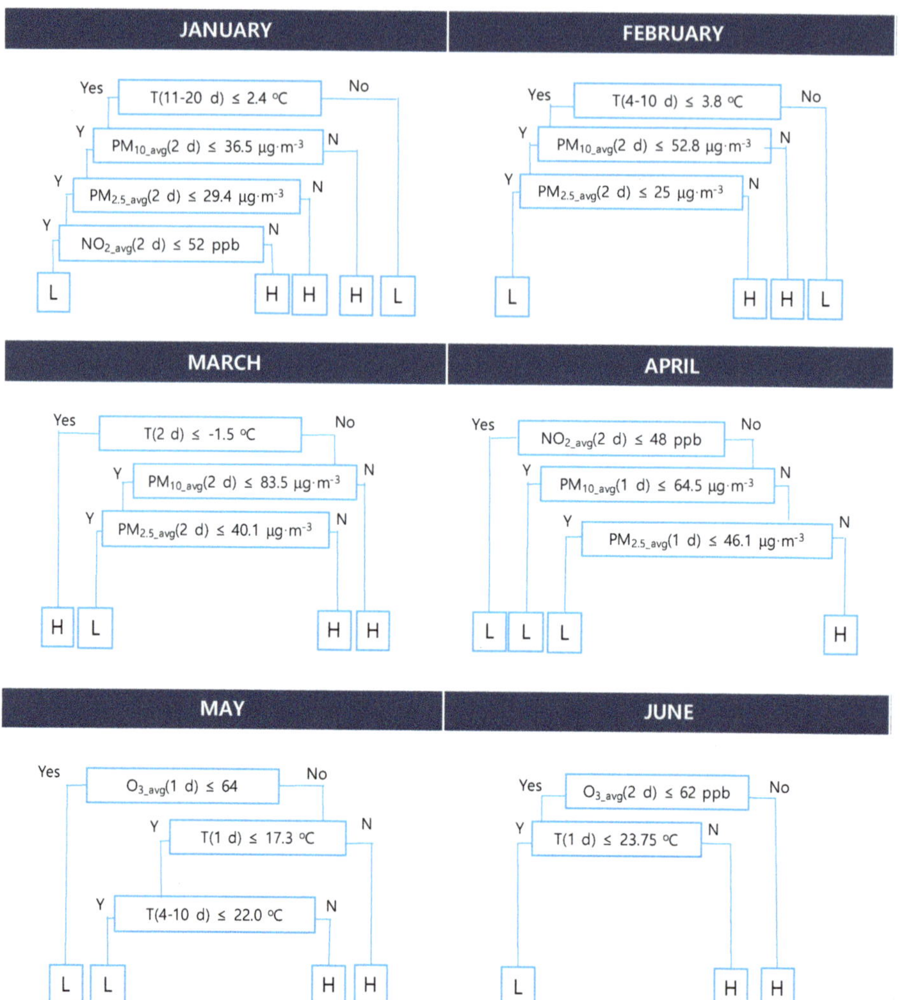

Figure 9. *Cont.*

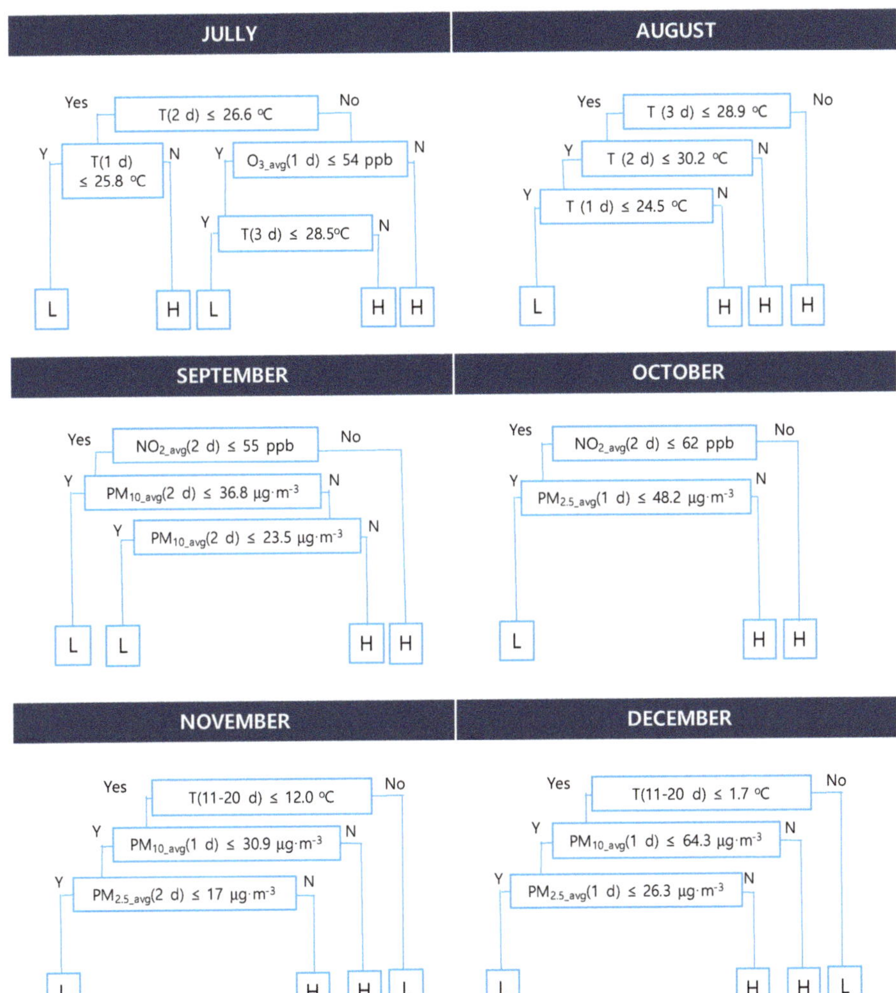

Figure 9. Daily average pollutant concentrations and temperatures affecting the high (H) and low (L) probability of deaths from respiratory disease.

Table 2. Ten-fold cross validation errors from using a decision tree model to predict higher or lower probability of death from respiratory disease.

Month	Model is Constructed Using: 1-h Maximum Pollutant Concentration (A)	Model is Constructed Using: 24-h Average Pollutant Concentration (B)	Difference (A)–(B)
January	30.6% (33.8%) [1]	28.8% (32.4%) [1]	1.8%
February	32.9% (31.4%) [1]	29.1% (31.4%) [1]	3.8%
March	28.7% (29.8%) [1]	25.1% (22.2%) [1]	3.6%
April	27.6% (31.6%) [1]	26.5% (24.5%) [1]	1.1%
May	28.0% (26.9%) [1]	29.9% (28.2%) [1]	−1.9%
June	32.2% (29.4%) [1]	33.2% (32.4%) [1]	−1.0%
July	38.0% (40.0%) [1]	37.9% (42.2%) [1]	0.1%
August	42.0% (41.0%) [1]	42.0% (41.0%) [1]	0.0%
September	38.8% (41.1%) [1]	36.3% (40.8%) [1]	2.5%
October	36.0% (36.3%) [1]	34.1% (35.6%) [1]	1.9%
November	34.0% (35.2%) [1]	29.5% (34.0%) [1]	4.5%
December	30.5% (33.1%) [1]	28.4% (32.2%) [1]	2.1%

[1] Values in parenthesis are errors using 3-day cumulative data instead of using data on each day.

5. Conclusions

The impact of air pollution on the risk of death from respiratory disease was analyzed. The analysis was conducted separately for each month to take into account the seasonal variability of air pollutant concentrations. The independent variables were 1-h maximum and 24-h average O_3, $PM_{2.5}$, PM_{10}, NO_2, SO_2, and CO concentrations and temperatures. The dependent variables were classified into days with high (H) and low (L) numbers of deaths caused by respiratory disease. The results showed that a higher risk of mortality was observed on days from November to March, in which $PM_{10}/PM_{2.5}$ concentrations were relatively high and temperatures were low. NO_2 was a crucial factor that influenced deaths from April to October. O_3 was the most important factor in May and in June. The risk of death increased with the high temperatures in July and August.

The accuracy of results was validated through 10-fold cross validation errors. Although error rates using the 1-h maximum pollutant concentrations were 27.6–42%, those using the 24-h average pollutant concentrations slightly decreased to 24.9–42%. Thus, the 24-h average pollutant concentrations were found to be more directly related with mortality from respiratory disease.

The results obtained from this study may be used to establish policies to minimize the adverse health effects of air pollution. For example, when pollutant concentrations are forecast and are communicated to the public, daily average concentrations should be emphasized more than the hourly maximum concentrations. In addition, the results could be used to guide the public to refrain from outdoor activities when pollutant levels are elevated.

Conflicts of Interest: The author declares no conflict of interest.

Appendix A

Basic statistics of hourly SO_2, CO, O_3, NO_2, PM_{10}, and $PM_{2.5}$ concentrations in Seoul, Korea, from 2008 and 2017 are illustrated as a box plot (Figure A1). The top and bottom of the box represented the 75th percentile (Q_3) and 25th percentile (Q_1), respectively. The tail's upper most value expressed the smaller one between the maximum value and $Q_3 + 1.5 \times (Q_3 - Q_1)$. On the other hand, the tail's lower most value is the larger one between the minimum value and $Q_1 - 1.5 \times (Q_3 - Q_1)$.

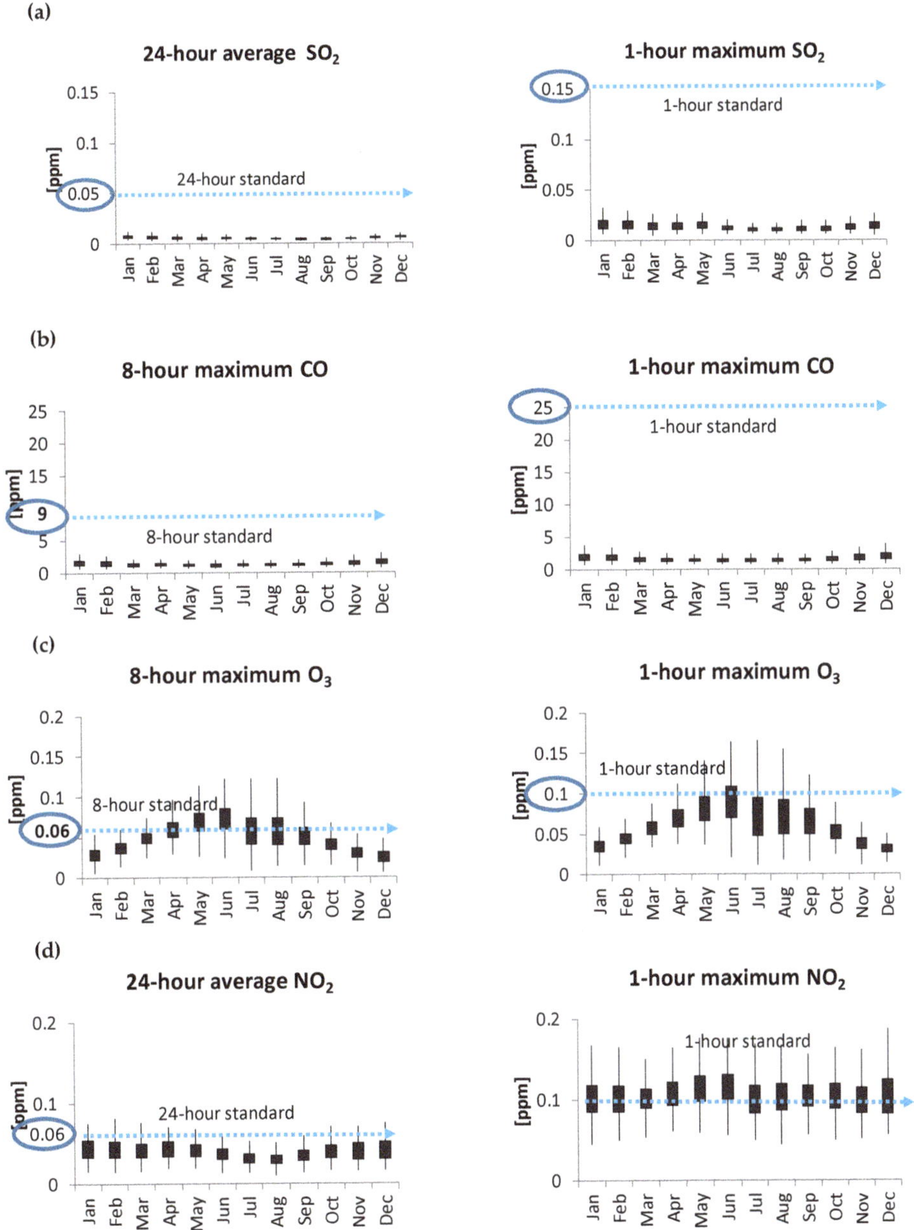

Figure A1. *Cont.*

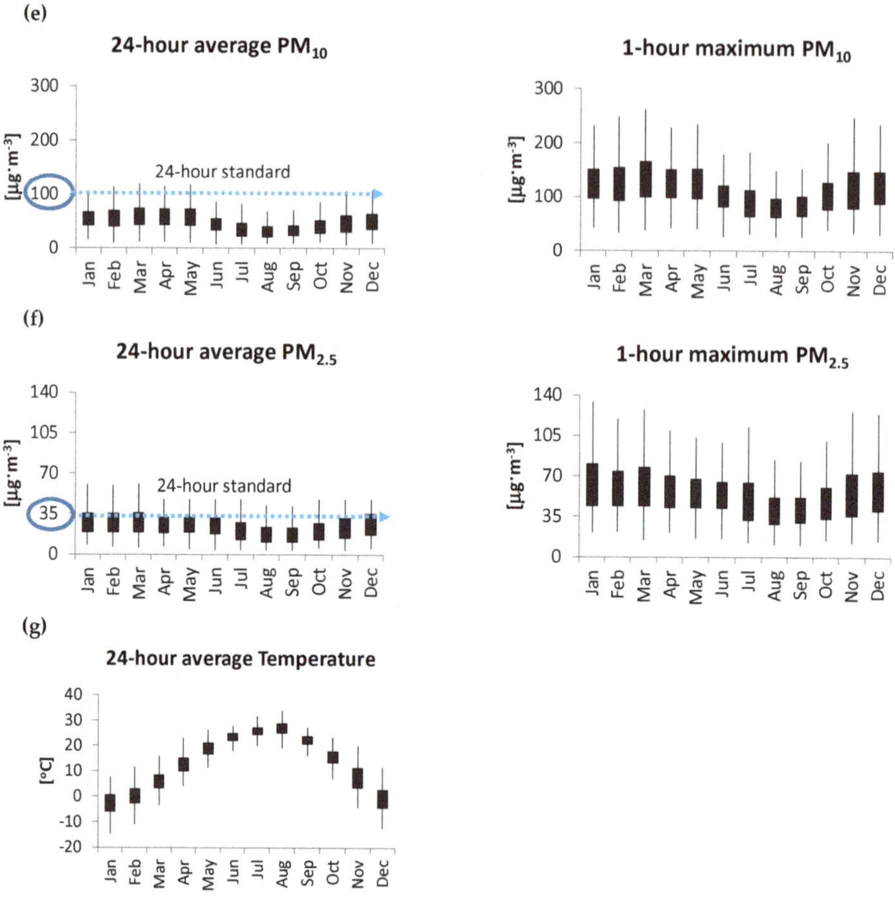

Figure A1. Box plot of 1-h maximum, 8-h maximum, and 24-h average pollutant concentrations, and 24-h average temperatures between 2008 and 2017. (**a**) SO_2, (**b**) CO, (**c**) O_3, (**d**) NO_2, (**e**) PM_{10} and (**f**) $PM_{2.5}$, and (**g**) temperature.

References

1. IPCC. *Chapter 11. Human Health: Impacts, Adaptation, and Co-Benefits. In AR5 Climate Change 2014: Impacts, Adaptation, and Vulnerability*; Working Group II contribution to the IPCC's Fifth Assessment Repport; Intergovernmental Panel on Climage Change; Cambridge University Press: New York, NY, USA, 2014.
2. Gao, F.; Calatayud, V.; García-Breijo, F.; Reig-Armiñana, J.; Feng, Z. Effects of elevated ozone on physiological, anatomical and ultrastructural characteristics of four common urban tree species in China. *Ecol. Indic.* **2016**, *67*, 367–379. [CrossRef]
3. Wang, L.; He, X.; Chen, W. Effects of elevated ozone on photosynthetic CO2 exchange and chlorophyll a fluorescence in leaves of quercus mongolica grown in urban area. *Bull. Environ. Contam. Toxicol.* **2009**, *82*, 478–481. [CrossRef] [PubMed]
4. Wu, Y.-L.; Lin, C.-H.; Lai, C.-H.; Lai, H.-C.; Young, C.-Y. Effects of Local Circulations, Turbulent Internal Boundary Layers, and Elevated Industrial Plumes on Coastal Ozone Pollution in the Downwind Kaohsiung Urban-Industrial Complex. *Terr. Atmos. Ocean. Sci.* **2010**, *21*, 343–357. [CrossRef]
5. Bell, M.L.; McDermott, A.; Zeger, S.L.; Samet, J.M.; Dominici, F. Ozone and short-term mortality in 95 US urban communities, 1987–2000. *JAMA* **2004**, *292*, 2372–2378. [CrossRef]

6. Reid, C.E.; Snowden, J.M.; Kontgis, C.; Tager, I.B. The Role of Ambient Ozone in Epidemiologic Studies of Heat-Related Mortality. *Environ. Health Perspect.* **2012**, *120*, 1627–1630. [CrossRef] [PubMed]
7. Stafoggia, M.; Forastiere, F.; Faustini, A.; Biggeri, A.; Bisanti, L.; Cadum, E.; Cernigliaro, A.; Mallone, S.; Pandolfi, P.; Serinelli, M.; et al. Susceptibility factors to ozone-related mortality: A population-based case-crossover analysis. *Am. J. Respir. Crit. Care Med.* **2010**, *182*, 376–384. [CrossRef]
8. Seven Million Premature Deaths Annually Linked to Air Pollution. Available online: https://www.who.int/mediacentre/news/releases/2014/air-pollution/en/ (accessed on 2 October 2019).
9. Hong, Y.-C.; Lee, J.-T.; Kim, H.; Kwon, H.-J. Air pollution: A new risk factor in ischemic stroke mortality. *Stroke* **2002**, *33*, 2165–2169. [CrossRef]
10. Jackson, J.E.; Yost, M.G.; Karr, C.; Fitzpatrick, C.; Lamb, B.K.; Chung, S.H.; Chen, J.; Avise, J.; Rosenblatt, R.A.; Fenske, R.A. Public health impacts of climate change in Washington State: Projected mortality risks due to heat events and air pollution. *Clim. Chang.* **2010**, *102*, 159–186. [CrossRef]
11. Son, J.-Y.; Bell, M.L.; Lee, J.-T. Individual exposure to air pollution and lung function in Korea: Spatial analysis using multiple exposure approaches. *Environ. Res.* **2010**, *110*, 739–749. [CrossRef]
12. Forns, J.; Dadvand, P.; Esnaola, M.; Alvarez-Pedrerol, M.; López-Vicente, M.; Garcia-Esteban, R.; Cirach, M.; Basagaña, X.; Guxens, M.; Sunyer, J. Longitudinal association between air pollution exposure at school and cognitive development in school children over a period of 3.5 years. *Environ. Res.* **2017**, *159*, 416–421. [CrossRef]
13. Ljungman, P.L.; Li, W.; Rice, M.B.; Wilker, E.H.; Schwartz, J.; Gold, D.R.; Koutrakis, P.; Benjamin, E.J.; Vasan, R.S.; Mitchell, G.F.; et al. Long- and short-term air pollution exposure and measures of arterial stiffness in the Framingham Heart Study. *Environ. Int.* **2018**, *121*, 139–147. [CrossRef] [PubMed]
14. Lee, H.; Myung, W.; Jeong, B.-H.; Choi, H.; Jhun, B.W.; Kim, H. Short- and long-term exposure to ambient air pollution and circulating biomarkers of inflammation in non-smokers: A hospital-based cohort study in South Korea. *Environ. Int.* **2018**, *119*, 264–273. [CrossRef] [PubMed]
15. Breiman, L.; Friedman, J.; Stone, C.J.; Olshen, R.A. *Classification and Regression Trees (Wadsworth Statistics/Probability)*, 1st ed.; Chapman and Hall/CRC: Boca Raton, FL, USA, 1984.
16. Air Quality Data. Available online: https://www.airkorea.or.kr/eng (accessed on 2 October 2019).
17. Synoptic Meteorological Data. Available online: https://data.kma.go.kr (accessed on 2 October 2019).
18. Microdata Integrated Service. Available online: https://mdis.kostat.go.kr (accessed on 2 October 2019).
19. Hastie, T.; Tibshirani, R.; Friedman, J. *The Elements of Statistical Learning*; Springer Science and Business Media LLC: New York, NY, USA, 2009.
20. Venkatasubramaniam, A.; Wolfson, J.; Mitchell, N.; Barnes, T.; Jaka, M.; French, S. Decision trees in epidemiological research. *Emerg. Themes Epidemiol.* **2017**, *14*, 11. [CrossRef]
21. Wolfson, J.; Venkatasubramaniam, A. Branching Out: Use of Decision Trees in Epidemiology. *Curr. Epidemiol. Rep.* **2018**, *5*, 221–229. [CrossRef]
22. Bae, J.-M. Clinical Decision Analysis using Decision Tree. *Epidemiol. Health* **2014**, *36*, 201425. [CrossRef] [PubMed]
23. Le Ray, I.; Lee, B.; Wikman, A.; Reilly, M. Evaluation of a decision tree for efficient antenatal red blood cell antibody screening. *Epidemiology* **2018**, *29*, 453–457. [CrossRef]
24. Chu, H.-J.; Lin, C.-Y.; Liau, C.-J.; Kuo, Y.-M. Identifying controlling factors of ground-level ozone levels over southwestern Taiwan using a decision tree. *Atmos. Environ.* **2012**, *60*, 142–152. [CrossRef]
25. Gardner, M.; Dorling, S. Statistical surface ozone models: An improved methodology to account for non-linear behaviour. *Atmos. Environ.* **2000**, *34*, 21–34. [CrossRef]
26. Park, S.-K. Assessing Factors Linked with Ozone Exceedances in Seoul, Korea through a Decision Tree Algorithm. *J. Environ. Sci. Int.* **2016**, *25*, 191–216. [CrossRef]
27. Baranski, L.; Jankowiak, L.; Rozemski, K. Seasonal changes of the spatial distribution of ozone content over Central Europe from TOVS/NOAA satellite data. *Adv. Space Res.* **1993**, *13*, 325–328. [CrossRef]
28. Roy, S.; Beig, G.; Jacob, D. Seasonal distribution of ozone and its precursors over the tropical Indian region using regional chemistry-transport model. *J. Geophys. Res. Space Phys.* **2008**, *113*, 21307. [CrossRef]
29. Scheel, H.E.; Areskoug, H.; Geiss, H.; Gomiscek, B.; Granby, K.; Haszpra, L.; Klasinc, L.; Kley, D.; Laurila, T.; Lindskog, A.; et al. On the Spatial Distribution and Seasonal Variation of Lower-Troposphere Ozone over Europe. *J. Atmos. Chem.* **1997**, *28*, 11–28. [CrossRef]

30. Kim, S.-W.; Yoon, S.-C.; Won, J.-G.; Choi, S.-C. Ground-based remote sensing measurements of aerosol and ozone in an urban area: A case study of mixing height evolution and its effect on ground-level ozone concentrations. *Atmos. Environ.* **2007**, *41*, 7069–7081. [CrossRef]
31. De Leon, S.F.; Thurston, G.D.; Ito, K. Contribution of Respiratory Disease to Nonrespiratory Mortality Associations with Air Pollution. *Am. J. Respir. Crit. Care Med.* **2003**, *167*, 1117–1123. [CrossRef] [PubMed]
32. Dabass, A.; Talbott, E.O.; Venkat, A.; Rager, J.; Marsh, G.M.; Sharma, R.K.; Holguin, F. Association of exposure to particulate matter (PM2.5) air pollution and biomarkers of cardiovascular disease risk in adult NHANES participants (2001–2008). *Int. J. Hyg. Environ. Health* **2016**, *219*, 301–310. [CrossRef] [PubMed]
33. Nath, S.; Patil, R.S. Prediction of air pollution concentration using an in situ real time mixing height model. *Atmos. Environ.* **2006**, *40*, 3816–3822. [CrossRef]
34. Han, Y.-J.; Holsen, T.M.; Hopke, P.K.; Cheong, J.-P.; Kim, H.; Yi, S.-M. Identification of source locations for atmospheric dry deposition of heavy metals during yellow-sand events in Seoul, Korea in 1998 using hybrid receptor models. *Atmos. Environ.* **2004**, *38*, 5353–5361. [CrossRef]
35. Gómez-Acebo, I.; Llorca, J.; Dierssen, T. Cold-related mortality due to cardiovascular diseases, respiratory diseases and cancer: A case-crossover study. *Public Health* **2012**, *127*, 252–258. [CrossRef]
36. Kim, D.W.; Deo, R.C.; Chung, J.H.; Lee, J.S. Projection of heat wave mortality related to climate change in Korea. *Nat. Hazards* **2016**, *80*, 623–637. [CrossRef]
37. Hien, P.; Loc, P.; Dao, N. Air pollution episodes associated with East Asian winter monsoons. *Sci. Total. Environ.* **2011**, *409*, 5063–5068. [CrossRef]
38. Park, S.K. Assessing the impact of air pollution on mortality rate from cardiovascular disease in Seoul, Korea. *Environ. Eng. Res.* **2018**, *23*, 430–441. [CrossRef]
39. Kim, K.; Macdonald, A.M.; Park, K.; Kim, H.-C.; Yoon, H.-I.; Yang, E.J.; Jung, J.; Kim, J.-H.; Lee, J.; Choi, T.; et al. Spatial and temporal variabilities of spring Asian dust events and their impacts on chlorophyll-a concentrations in the western North Pacific Ocean. *Geophys. Res. Lett.* **2017**, *44*, 1474–1482.
40. Park, S.-U.; Choe, A.; Park, M.-S. A simulation of Asian dust events observed from 20 to 29 December 2009 in Korea by using ADAM2. *Asia-Pacific J. Atmos. Sci.* **2013**, *49*, 95–109. [CrossRef]
41. Dong, G.-H.; Zhang, P.; Sun, B.; Zhang, L.; Chen, X.; Ma, N.; Yu, F.; Guo, H.; Huang, H.; Lee, Y.L.; et al. Long-Term Exposure to Ambient Air Pollution and Respiratory Disease Mortality in Shenyang, China: A 12-Year Population-Based Retrospective Cohort Study. *Respiration* **2012**, *84*, 360–368. [CrossRef]
42. Brook, J.R.; Yung, W.T.; Dales, R.E.; Burnett, R.T.; Krewski, D. Association between Ozone and Hospitalization for Respiratory Diseases in 16 Canadian Cities. *Environ. Res.* **1997**, *72*, 24–31.
43. Lee, W.; Choi, H.M.; Kim, D.; Honda, Y.; Guo, Y.L.; Kim, H. Temporal changes in morality attributed to heat extremes for 57 cities in Northeast Asia. *Sci. Total Environ.* **2018**, *616*, 703–709. [CrossRef]
44. Camacho, J.; Ferrer, A.; Páez, J.C. Cross-validation in PCA models with the element-wise k-fold (ekf) algorithm: Practical aspects. *Chemom. Intell. Lab. Syst.* **2014**, *131*, 37–50. [CrossRef]
45. Grimm, K.J.; Mazza, G.L.; Davoudzadeh, P. Model Selection in Finite Mixture Models: A k-Fold Cross-Validation Approach. *Struct. Equ. Model.* **2017**, *24*, 246–256. [CrossRef]
46. Lim, C.; Jang, J.; Lee, I.; Kim, G.; Lee, S.M.; Kim, Y.; Kim, H.; Kaufman, A.J. Sulfur isotope and chemical compositions of the wet precipitation in two major urban areas, Seoul and Busan, Korea. *J. Asian Earth Sci.* **2014**, *79*, 415–425. [CrossRef]
47. Wang, B.; Ding, Q.; Jhun, J.G. Trends in Seoul (1778–2004) summer precipitation. *Geophys. Res. Lett.* **2006**, *33*, L15803. [CrossRef]
48. Zanobetti, A.; Schwartz, J.; Samoli, E.; Gryparis, A.; Touloumi, G.; Peacock, J.; Anderson, R.H.; Le Tertre, A.; Bobros, J.; Celko, M.; et al. The temporal pattern of respiratory and heart disease mortality in response to air pollution. *Environ. Health Perspect.* **2003**, *111*, 1188–1193. [CrossRef] [PubMed]
49. Jerrett, M.; Burnett, R.T.; Brook, J.; Kanaroglou, P.; Giovis, C.; Finkelstein, N.; Hutchison, B. Do socioeconomic characteristics modify the short term association between air pollution and mortality? Evidence from a zonal time series in Hamilton, Canada. *J. Epidemiol. Community Health* **2004**, *58*, 31–40. [CrossRef] [PubMed]

MDPI
St. Alban-Anlage 66
4052 Basel
Switzerland
Tel. +41 61 683 77 34
Fax +41 61 302 89 18
www.mdpi.com

Atmosphere Editorial Office
E-mail: atmosphere@mdpi.com
www.mdpi.com/journal/atmosphere

www.ingramcontent.com/pod-product-compliance
Lightning Source LLC
LaVergne TN
LVHW070905170726
843515LV00005B/710

* 9 7 8 3 0 3 9 3 6 8 2 8 0 *